I0055877

FIELD GUIDE
TO
CONTINUOUS
PROBABILITY DISTRIBUTIONS

Gavin E. Crooks

v 1.0.0

2019

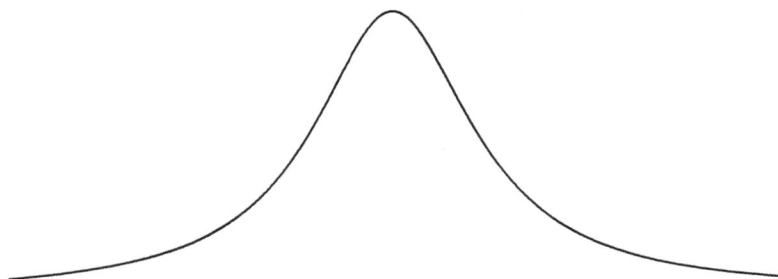

v 1.0.0

Copyright © 2010-2019 Gavin E. Crooks

ISBN: 978-1-7339381-0-5

http://threeplusone.com/fieldguide
Berkeley Institute for Theoretical Sciences (BITS)
typeset on 2019-04-08 with XeTeX version 0.99999
fonts: Trump Mediaeval (text), Euler (math)
2 7 1 8 2 8 1 8 2 8 4 5 9 0 4

Preface: The search for GUD

A common problem is that of describing the probability distribution of a single, continuous variable. A few distributions, such as the normal and exponential, were discovered in the 1800's or earlier. But about a century ago the great statistician, Karl Pearson, realized that the known probability distributions were not sufficient to handle all of the phenomena then under investigation, and set out to create new distributions with useful properties.

During the 20th century this process continued with abandon and a vast menagerie of distinct mathematical forms were discovered and invented, investigated, analyzed, rediscovered and renamed, all for the purpose of describing the probability of some interesting variable. There are hundreds of named distributions and synonyms in current usage. The apparent diversity is unending and disorienting.

Fortunately, the situation is less confused than it might at first appear. Most common, continuous, univariate, unimodal distributions can be organized into a small number of distinct families, which are all special cases of a single Grand Unified Distribution. This compendium details these hundred or so simple distributions, their properties and their interrelations.

Gavin E. Crooks

Acknowledgments

In curating this collection of distributions, I have benefited greatly from Johnson, Kotz, and Balakrishnan's monumental compendiums [2, 3], Eric Weisstein's MathWorld, the Leemis chart of Univariate Distribution Relationships [8, 9], and myriad pseudo-anonymous contributors to Wikipedia. Additional contributions are noted in the version history below.

Version History

1.0 (2019-04-01) First print edition. Over 170 named univariate continuous probability distributions, and at least as many synonyms. Added inverse Maxwell (11.21), inverse half-normal (11.22), inverse Nakagami (11.23), reciprocal inverse Gaussian (20.4), generalized Sichel (20.14), Pearson exponential (20.15), Perks (20.16), and noncentral chi (21.14). distributions. Added diagram of the Pearson-exponential hierarchy (Fig. 3). Renamed the Pearson II distribution to central-beta, and the symmetric beta-logistic distribution to central-logistic.

0.12 (2019-02-23) Added Porter-Thomas (7.5), Epanechnikov (12.9), biweight (12.10), triweight (12.11), Libby-Novick (20.10), Gauss hypergeometric (20.11), confluent hypergeometric (20.12), Johnson S_u (21.10), and log-Cauchy (21.12) distributions.

0.11 (2017-06-19) Added hyperbola (20.6), Halphen (20.5), Halphen B (20.7), inverse Halphen B (20.8), generalized Halphen (20.13), Sichel (20.9) and Appell Beta (20.17) distributions. Thanks to Saralees Nadarajah.

0.10 (2017-02-08) Added K (21.8) and generalized K (21.5) distributions. Clarified notation and nomenclature. Thanks to Harish Vangala.

0.9 (2016-10-18) Added pseudo-Voigt (21.17), and Student's t_3 (9.4) distributions. Reparameterized hyperbolic sine (14.3) distribution. Renamed inverse Burr to Dagum (18.4). Derived limit of unit gamma to log-normal (p68). Corrected spelling of "arrises" (sharp edges formed by the meeting of surfaces) to "arises" (emerge; become apparent).

0.8 (2016-08-30) The Unprincipled edition: Added Moyal distribution (8.8), a special case of the gamma-exponential distribution. Corrected spelling of "principle" to "principal". Thanks to Matthew Hankins and Mara Averick.

0.7 (2016-04-05) Added Hohlfeld distribution. Added appendix on limits. Reformatted and rationalized distribution hierarchy diagrams. Thanks to Phill Geissler.

0.6 (2014-12-22) Added appendix on the algebra of random variables. Added Box-Muller transformation. For the gamma-exponential distribution, switched the sign on the parameter α. Fixed the relation between beta distributions and ratios of gamma distributions (α and γ were switched in most cases). Thanks to Fabian Krüger and Lawrence Leemis.

0.5 (2013-07-01) Added uniform product, half generalized Pearson VII, half exponential power, GUD and q-type distributions. Moved Pearson IV to own section. Fixed errors in inverse Gaussian. Added random variate generation to appendix. Thanks to David Sivak, Dieter Grientschnig, Srividya Iyer-Biswas, and Shervin Fatehi.

0.4 (2012-03-01) Added erratics. Moved gamma distribution to own section. Renamed log-gamma to gamma-exponential. Added permalink. Added new tree of distributions. Thanks to David Sivak and Frederik Beaujean.

0.3 (2011-06-40) Added tree of distributions.

0.2 (2011-03-01) Expanded families. Thanks to David Sivak.

0.1 (2011-01-16) Initial release. Organize over 100 simple, continuous, univariate probability distributions into 14 families. Greatly expands on previous paper that discussed the Amoroso and gamma-exponential families [10]. Thanks to David Sivak, Edward E. Ayoub, and Francis J. O'Brien.

Endorsements

"Ridiculously useful!" – Mara Averick[1]

"Abramowitz and Stegun for probability distributions"– Kranthi K. Mandadapu[2]

"Awesome" – Avery Brooks[3]

"Who are you? How did you get in my house?" – Donald Knuth[4]

---

[1] twitter
[2] Thursday Lunch with Scientists
[3] Private communication
[4] https://xkcd.com/163/

Contents

One shape parameter

Two shape parameters

Three (or more) shape parameters

Miscellanea

Appendix

Contents

List of Figures

LIST OF TABLES

Figure 1: Hierarchy of principal distributions

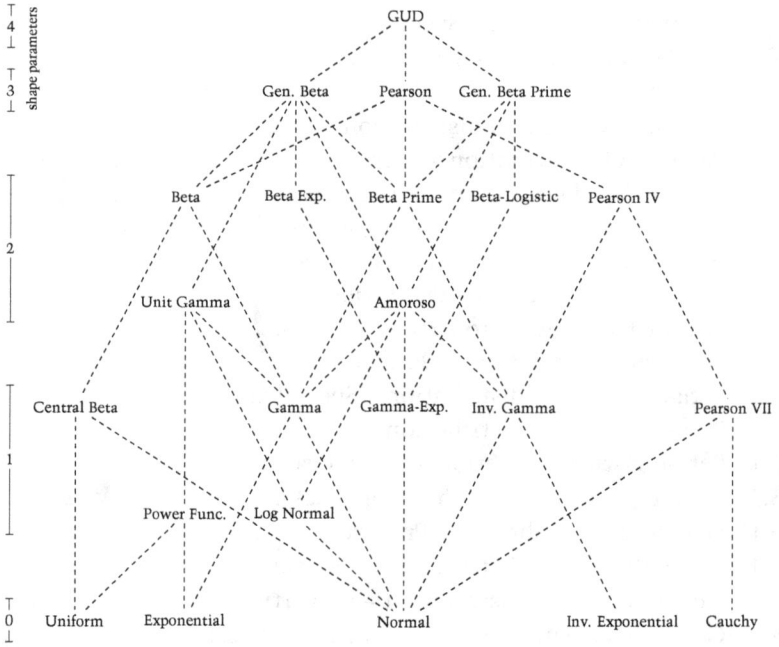

Figure 2: Hierarchy of Pearson distributions

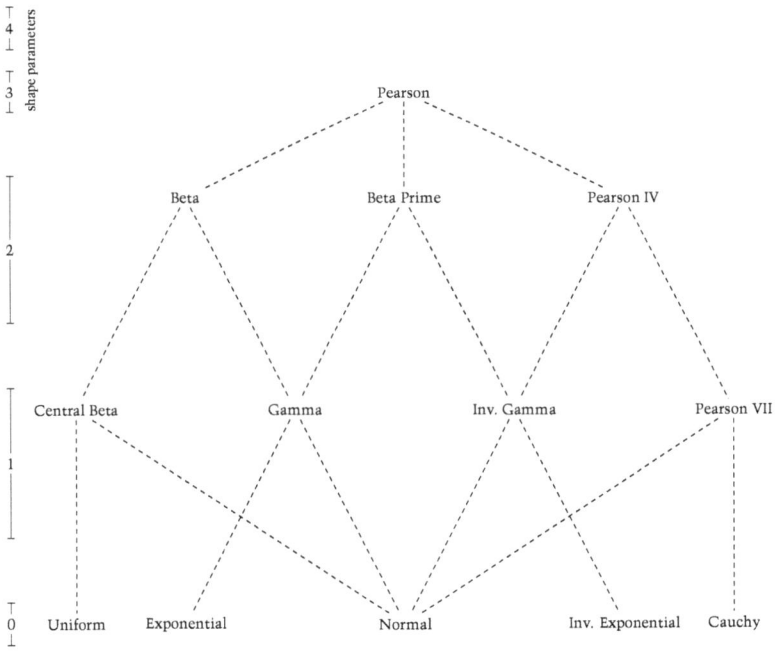

Figure 3: Hierarchy of Pearson-exponential distributions

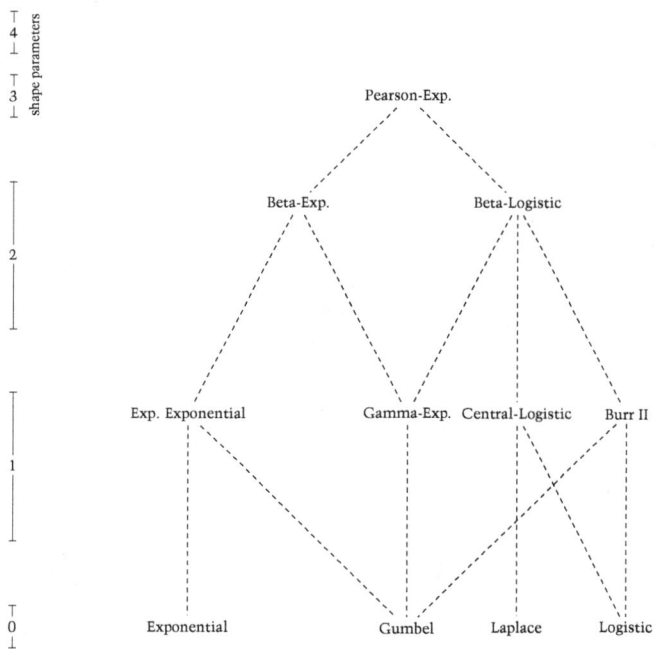

Figure 4: Hierarchy of extreme order statistics

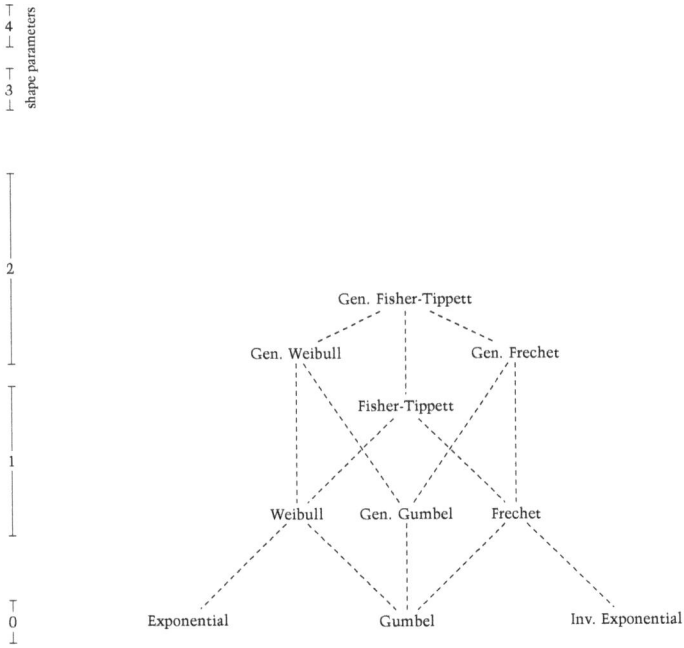

Figure 5: Hierarchies of symmetric simple distributions

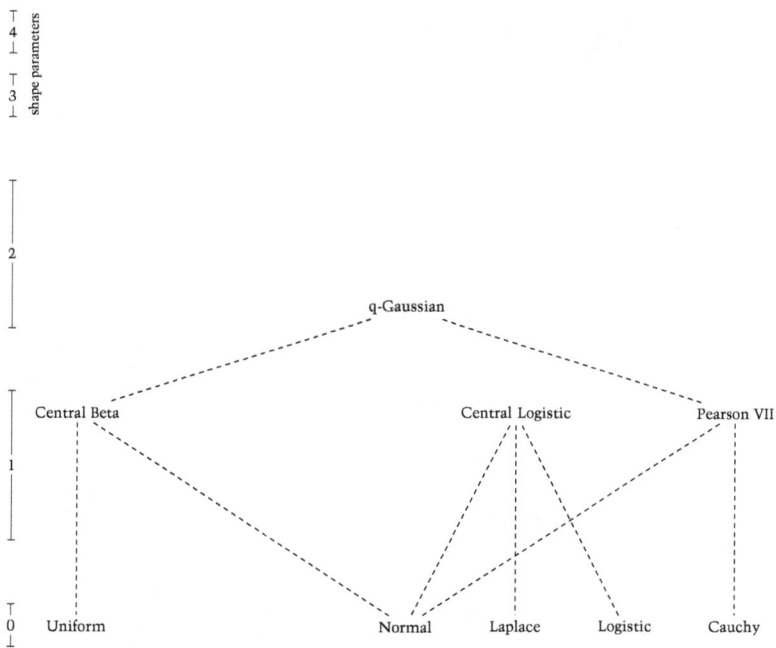

1   UNIFORM DISTRIBUTION

The simplest continuous distribution is a uniform density over an interval.

Uniform (flat, rectangular) distribution:

$$\text{Uniform}(x\ ;\ a, s) = \frac{1}{|s|} \tag{1.1}$$

$$\text{for } a, s \text{ in } \mathbb{R},$$

$$\text{support } x \in [a, a+s], \quad s > 0$$

$$x \in [a+s, a], \quad s < 0$$

The uniform distribution is also commonly parameterized with the boundary points, a and $b = a + s$, rather than location a and scale s as here. Note that the discrete analog of the continuous uniform distribution is also often referred to as the uniform distribution.

Special cases

The **standard uniform** distribution covers the unit interval, $x \in [0, 1]$.

$$\text{StdUniform}(x) = \text{Uniform}(x\ ;\ 0, 1) \tag{1.2}$$

The **standardized uniform** distribution, with zero mean and unit variance, is $\text{Uniform}(x\ ;\ -\sqrt{3}, 2\sqrt{3})$.

Three limits of the uniform distribution are important. If one of the boundary points is infinite (infinite scale), then we obtain an improper (unnormalizable) **half-uniform** distribution. In the limit that both boundary points reach infinity (with opposite signs) we obtain an **unbounded uniform** distribution. In the alternative limit that the boundary points converge, we obtain a **degenerate** (delta, Dirac) distribution, wherein the entire probability density is concentrated on a single point.

Interrelations

Uniform distributions, with finite, semi-infinite, or infinite support, are limits of many distribution families. The finite uniform distribution is a

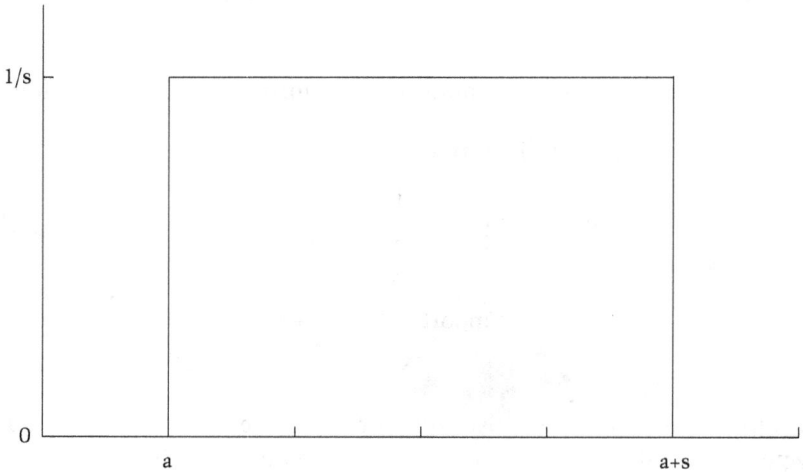

Figure 6: Uniform distribution, Uniform$(x\ ;\ a, s)$ (1.1)

special case of the beta distribution (12.1).

$$\text{Uniform}(x\ ;\ a, s) = \text{Beta}(x\ ;\ a, s, 1, 1)$$
$$= \text{CentralBeta}(x\ ;\ a + \tfrac{s}{2}, s)$$

Similarly, the semi-infinite uniform distribution is a limit of the Pareto (5.5), beta prime (13.1), Amoroso (11.1), gamma (7.1), and exponential (2.1) distributions, and the infinite support uniform distribution is a limit of the normal (4.1), Cauchy (9.6), logistic (15.5) and gamma-exponential (8.1) distributions, among others.

The order statistics (§C) of the uniform distribution is the beta distribution (12.1).

$$\text{OrderStatistic}_{\text{Uniform}(a,s)}(x\ ;\ \alpha, \gamma) = \text{Beta}(x\ ;\ a, s, \alpha, \gamma)$$

The standard uniform distribution is related to every other continuous distribution via the inverse probability integral transform (Smirnov transform). If X is a random variable and $F_X^{-1}(z)$ is the inverse of the corresponding cumulative distribution function then

$$X \sim F_X^{-1}\left(\text{StdUniform}()\right) .$$

If the inverse cumulative distribution function has a tractable closed form, then inverse transform sampling can provide an efficient method of sampling random numbers from the distribution of interest. See appendix (§E).

The power function distribution (5.1) is related to the uniform distribution via a Weibull transform.

$$\text{PowerFn}(a, s, \beta) \sim a + s \; \text{StdUniform}()^{\frac{1}{\beta}}$$

The sum of n independent standard uniform variates is the Irwin-Hall (21.9) distribution,

$$\sum_{i=1}^{n} \text{Uniform}_i(0, 1) \sim \text{IrwinHall}(n)$$

and the product is the uniform-product distribution (10.2).

$$\prod_{i=1}^{n} \text{Uniform}_i(0, 1) \sim \text{UniformProduct}(n)$$

Table 1.1: Properties of the uniform distribution

Properties		
notation	$\mathrm{Uniform}(x \; ; a, s)$	
PDF	$\dfrac{1}{|s|}$	
CDF/CCDF	$\frac{x-a}{s}$	$s > 0 \,/\, s < 0$
parameters	$a, \; s$ in $\mathbb{R}$	
support	$a \leqslant x \leqslant a + s$	$s > 0$
	$a + s \leqslant x \leqslant a$	$s < 0$
median	$a + \frac{1}{2}s$	
mode	any supported value	
mean	$a + \frac{1}{2}s$	
variance	$\frac{1}{12}s^2$	
skew	0	
ex. kurtosis	$-\frac{6}{5}$	
entropy	$\ln|s|$	
MGF	$\dfrac{e^{at}(e^{st} - 1)}{|s|t}$	
CF	$\dfrac{e^{iat}(e^{ist}) - 1}{i|s|t}$	

2   EXPONENTIAL DISTRIBUTION

Exponential (Pearson type X, waiting time, negative exponential, inverse exponential) distribution [7, 11, 2]:

$$\text{Exp}(x \; ; a, \theta) = \frac{1}{|\theta|} \exp\left\{ -\frac{x - a}{\theta} \right\} \qquad (2.1)$$

$$a, \; \theta, \; \text{in} \; \mathbb{R}$$

$$\text{support } x > a, \quad \theta > 0$$

$$x < a, \quad \theta < 0$$

An important property of the exponential distribution is that it is memoryless: assuming positive scale and zero location ($a = 0$, $\theta > 0$) the conditional probability given that $x > c$, where c is a positive content, is again an exponential distribution with the same scale parameter. The only other distribution with this property is the geometric distribution, the discrete analog of the exponential distribution. The exponential is the maximum entropy distribution given the mean and semi-infinite support.

Special cases

The exponential distribution is commonly defined with zero location and positive scale (**anchored exponential**). With $a = 0$ and $\theta = 1$ we obtain the **standard exponential** distribution.

Interrelations

The exponential distribution is common limit of many distributions.

$$\text{Exp}(x \; ; a, \theta) = \text{Amoroso}(x \; ; a, \theta, 1, 1)$$

$$= \text{Gamma}(x \; ; a, \theta, 1)$$

$$\text{Exp}(x \; ; 0, \theta) = \text{Amoroso}(x \; ; 0, \theta, 1, 1)$$

$$= \text{Gamma}(x \; ; 0, \theta, 1)$$

$$\text{Exp}(x \; ; a, \theta) = \lim_{\beta \to \infty} \text{PowerFn}(x \; ; a - \beta\theta, \beta\theta, \beta)$$

The sum of independent exponentials is an Erlang distribution, a special

Table 2.1: Properties of the exponential distribution

Properties		
notation	$\mathrm{Exp}(x\,;a,\theta)$	
PDF	$\dfrac{1}{|\theta|}\exp\left\{-\dfrac{x-a}{\theta}\right\}$	
CDF/CCDF	$1-\exp\left\{-\dfrac{x-a}{\theta}\right\}$	$\theta>0\,/\,\theta<0$
parameters	$a,\ \theta,\ \text{in}\ \mathbb{R}$	
support	$[a,+\infty]$	$\theta>0$
	$[-\infty,a]$	$\theta<0$
median	$a+\theta\ln 2$	
mode	a	
mean	$a+\theta$	
variance	θ^2	
skew	$\mathrm{sgn}(\theta)\,2$	
ex. kurtosis	6	
entropy	$1+\ln|\theta|$	
MGF	$\dfrac{\exp(at)}{(1-\theta t)}$	
CF	$\dfrac{\exp(iat)}{(1-i\theta t)}$	

case of the gamma distribution (7.1).

$$\sum_{i=1}^{n}\mathrm{Exp}_i(0,\theta)\sim\mathrm{Gamma}(0,\theta,n)$$

The minima of a collection of exponentials, with positive scales $\theta_i>0$, is also exponential,

$$\min\big(\mathrm{Exp}_1(0,\theta_1),\ \mathrm{Exp}_2(0,\theta_2),\ \ldots\ ,\ \mathrm{Exp}_n(0,\theta_n)\big)\sim\mathrm{Exp}(0,\theta')\,,$$

where $\theta'=(\sum_{i=1}^{n}\frac{1}{\theta_i})^{-1}$.

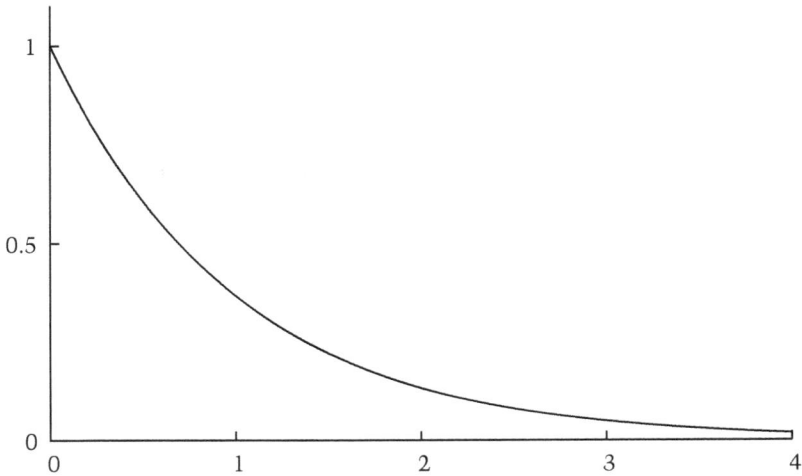

Figure 7: Standard exponential distribution, $\text{Exp}(x \,;\, 0, 1)$

The order statistics (§C) of the exponential distribution are the beta-exponential distribution (14.1).

$$\text{OrderStatistic}_{\text{Exp}(\zeta,\lambda)}(x \,;\, \alpha, \gamma) = \text{BetaExp}(x \,;\, \zeta, \lambda, \alpha, \gamma)$$

A Weibull transform of the standard exponential distribution yields the Weibull distribution (11.27).

$$\text{Weibull}(a, \theta, \beta) \sim a + \theta \; \text{StdExp}()^{\frac{1}{\beta}}$$

The ratio of independent anchored exponential distributions is the exponential ratio distribution (5.7), a special case of the beta prime distribution (13.1).

$$\text{BetaPrime}(0, \tfrac{\theta_1}{\theta_2}, 1, 1) \sim \text{ExpRatio}(0, \tfrac{\theta_1}{\theta_2}) \sim \frac{\text{Exp}_1(0, \theta_1)}{\text{Exp}_2(0, \theta_2)}$$

3 Laplace Distribution

Laplace (Laplacian, double exponential, Laplace's first law of error, two-tailed exponential, bilateral exponential, biexponential) distribution [12, 13, 14] is a two parameter, symmetric, continuous, univariate, unimodal probability density with infinite support, smooth expect for a single cusp. The functional form is

$$\text{Laplace}(x \; ; \zeta, \theta) = \frac{1}{2|\theta|} e^{-\left|\frac{x-\zeta}{\theta}\right|} \tag{3.1}$$

$$\text{for } x, \; \zeta, \; \theta \text{ in } \mathbb{R}$$

The two real parameters consist of a location parameter ζ, and a scale parameter θ.

Special cases

The **standard Laplace** (Poisson's first law of error) distribution occurs when $\zeta = 0$ and $\theta = 1$.

Interrelations

The Laplace distribution is a limit of the central-logistic (15.4), exponential power (21.4) and generalized Pearson VII (21.6) distributions.

As θ limits to infinity, the Laplace distribution limits to a degenerate distribution. In the alternative limit that θ limits to zero, we obtain an indefinite uniform distribution.

The difference between two independent identically distributed exponential random variables is Laplace, and therefore so is the time difference between two independent Poisson events.

$$\text{Laplace}(\zeta, \theta) \sim \text{Exp}_1(\zeta, \theta) - \text{Exp}_2(\zeta, \theta)$$

Conversely, the absolute value (about the centre of symmetry) is exponential.

$$\text{Exp}(\zeta, |\theta|) \sim |\text{Laplace}(\zeta, \theta) - \zeta| + \zeta$$

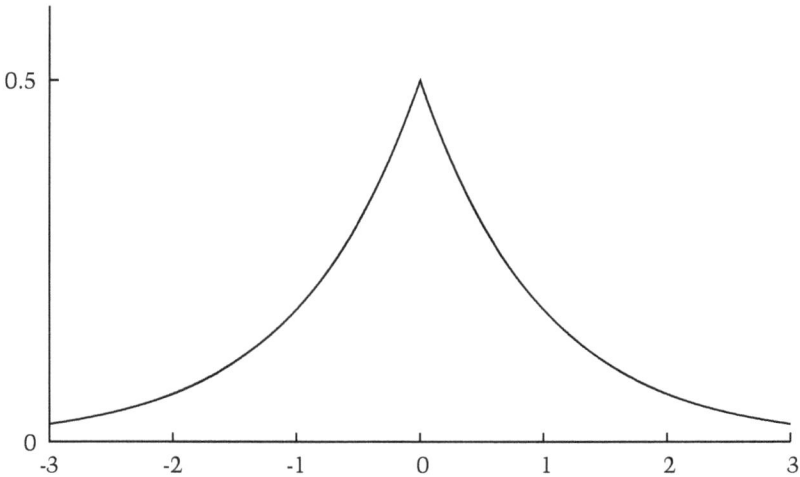

Figure 8: Standard Laplace distribution, Laplace$(x\ ;0,1)$

The log ratio of standard uniform distributions is a standard Laplace.

$$\text{Laplace}(0,1) \sim \ln \frac{\text{StdUniform}_1()}{\text{StdUniform}_2()}$$

The Fourier transform of a standard Laplace distribution is the standard Cauchy distribution (9.6).

$$\int_{-\infty}^{+\infty} \frac{1}{2}e^{-|x|}e^{itx}dx = \frac{1}{1+t^2}$$

Table 3.1: Properties of the Laplace distribution

Properties	
notation	$\mathrm{Laplace}(x \; ; \; \zeta, \theta)$
PDF	$\dfrac{1}{2|\theta|} e^{-|\frac{x-\zeta}{\theta}|}$
CDF	$\begin{cases} \frac{1}{2} e^{-|\frac{x-\zeta}{\theta}|} & x \leqslant \zeta \\ 1 - \frac{1}{2} e^{-|\frac{x-\zeta}{\theta}|} & x \geqslant \zeta \end{cases}$
parameters	ζ, θ in $\mathbb{R}$
support	$x \in [-\infty, +\infty]$
median	ζ
mode	ζ
mean	ζ
variance	$2\theta^2$
skew	0
ex. kurtosis	3
entropy	$1 + \ln(2|\theta|)$
MGF	$\dfrac{\exp(\zeta t)}{1 - \theta^2 t^2}$
CF	$\dfrac{\exp(i\zeta t)}{1 + \theta^2 t^2}$

4   NORMAL DISTRIBUTION

The **Normal** (Gauss, Gaussian, bell curve, Laplace-Gauss, de Moivre, error, Laplace's second law of error, law of error) [15, 2] distribution is a ubiquitous two parameter, continuous, univariate, unimodal probability distribution with infinite support, and an iconic bell shaped curve.

$$\text{Normal}(x \; ; \mu, \sigma) = \frac{1}{\sqrt{2\pi\sigma^2}} \exp\left\{-\frac{(x-\mu)^2}{2\sigma^2}\right\} \tag{4.1}$$

$$\text{for } x, \; \mu, \; \sigma \text{ in } \mathbb{R}$$

The location parameter μ is the mean, and the scale parameter σ is the standard deviation. Note that the normal distribution is often parameterized with the variance σ^2 rather than the standard deviation. Herein, we choose to consistently parameterize distributions with a scale parameter.

The normal distribution most often arises as a consequence of the famous central limit theorem, which states (in its simplest form) that the mean of independent and identically distribution random variables, with finite mean and variance, limit to the normal distribution as the sample size becomes large. The normal distribution is also the maximum entropy distribution for fixed mean and variance.

Special cases

With $\mu = 0$ and $\sigma = 1/\sqrt{2}h$ we obtain the **error function** distribution, and with $\mu = 0$ and $\sigma = 1$ we obtain the **standard normal** (Φ, z, unit normal) distribution.

Interrelations

In the limit that $\sigma \to \infty$ we obtain an unbounded uniform (flat) distribution, and in the limit $\sigma \to 0$ we obtain a degenerate (delta) distribution.

The normal distribution is a limiting form of many distributions, including the gamma-exponential (8.1), Amoroso (11.1) and Pearson IV (16.1) families and their superfamilies.

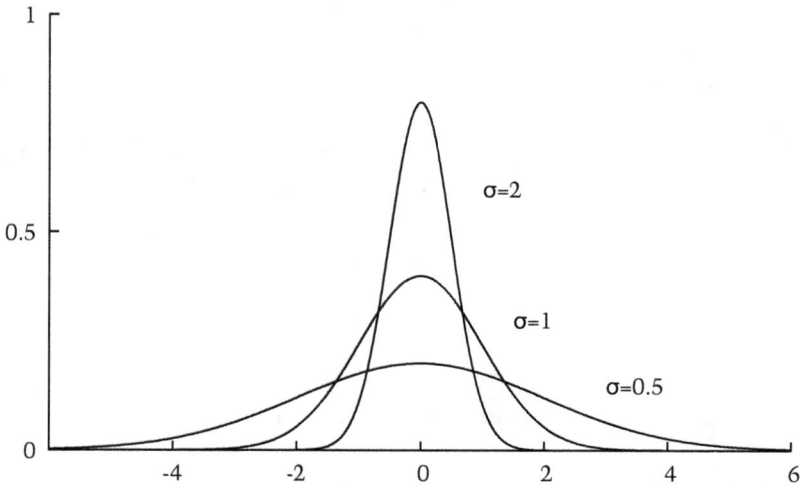

Figure 9: Normal distributions, $\text{Normal}(x \; ; 0, \sigma)$

Many distributions are transforms of normal distributions.

$$\exp\big(\text{Normal}(\mu, \sigma)\big) \sim \text{LogNormal}(0, e^\mu, \sigma) \qquad (6.1)$$

$$\big|\,\text{Normal}(0, \sigma)\big| \sim \text{HalfNormal}(\sigma) \qquad (11.7)$$

$$\text{StdNormal}()^2 \sim \text{ChiSqr}(1) \qquad (7.3)$$

$$\sum_{i=1,k} \text{StdNormal}_i()^2 \sim \text{ChiSqr}(k) \qquad (7.3)$$

$$\text{Normal}(0, \sigma)^{-2} \sim \text{Lévy}(0, \tfrac{1}{\sigma^2}) \qquad (11.15)$$

$$\big|\,\text{Normal}(0, \sigma)\big|^{\frac{2}{\beta}} \sim \text{Stacy}((2\sigma^2)^{\frac{1}{\beta}}, \tfrac{1}{2}, \beta) \qquad (11.2)$$

$$\frac{\text{StdNormal}_1()}{\text{StdNormal}_2()} \sim \text{StdCauchy}() \qquad (9.7)$$

The normal distribution is stable (21.20): A sum of independent normal random variables is also normally distributed.

$$\text{Normal}_1(\mu_1, \sigma_1) + \text{Normal}_2(\mu_2, \sigma_2) \sim \text{Normal}_3(\mu_1 + \mu_2, \sigma_1 + \sigma_2)$$

Table 4.1: Properties of the normal distribution

Properties	
notation	$\text{Normal}(x \; ; \mu, \sigma)$
PDF	$\dfrac{1}{\sqrt{2\pi\sigma^2}} \exp\left\{ -\dfrac{(x-\mu)^2}{2\sigma^2} \right\}$
CDF	$\dfrac{1}{2}\left[1 + \text{erf}\left(\dfrac{x-\mu}{\sqrt{2\sigma^2}} \right) \right]$
parameters	$\mu, \sigma \text{ in } \mathbb{R}$
support	$x \in [-\infty, +\infty]$
median	μ
mode	μ
mean	μ
variance	σ^2
skew	0
ex. kurtosis	0
entropy	$\frac{1}{2}\ln(2\pi e\sigma^2)$
MGF	$\exp\left(\mu t + \frac{1}{2}\sigma^2 t^2\right)$
CF	$\exp\left(i\mu t - \frac{1}{2}\sigma^2 t^2\right)$

The Box-Muller transform [16] generates pairs of independent normal variates from pairs of uniform random variates.

$$\text{StdNormal}_1() \sim \text{ChiSqr}(1) \; \cos\left(2\pi \; \text{StdUniform}_2()\right)$$
$$\text{StdNormal}_2() \sim \text{ChiSqr}(1) \; \sin\left(2\pi \; \text{StdUniform}_2()\right)$$
$$\text{where } \text{ChiSqr}(1) \sim \sqrt{-2\ln \text{StdUniform}_1()}$$

Nowadays more efficient random normal generation methods are generally employed (§E).

5 Power Function Distribution

Power function (power) distribution [7, 17, 3] is a three parameter, continuous, univariate, unimodal probability density, with finite or semi-infinite support. The functional form in the most straightforward parameterization consists of a single power function.

$$\text{PowerFn}(x \; ; a, s, \beta) = \left| \frac{\beta}{s} \right| \left(\frac{x-a}{s} \right)^{\beta-1} \tag{5.1}$$

$$\text{for } x, a, s, \beta \text{ in } \mathbb{R}$$

$$\text{support } x \in [a, a+s], s > 0, \; \beta > 0$$

$$\text{or } x \in [a+s, a], s < 0, \; \beta > 0$$

$$\text{or } x \in [a+s, +\infty], s > 0, \; \beta < 0$$

$$\text{or } x \in [-\infty, a+s], s < 0, \; \beta < 0$$

With positive β we obtain a distribution with finite support. But by allowing β to extend to negative numbers, the power function distribution also encompasses the semi-infinite Pareto distribution (5.5), and in the limit $\beta \to \infty$ the exponential distribution (2.1).

Alternative parameterizations

Generalized Pareto distribution: An alternative parameterization that emphasizes the limit to exponential.

$$\text{GenPareto}(x \; ; a', s', \xi) \tag{5.2}$$

$$= \begin{cases} \frac{1}{|\theta|} \left(1 + \xi \frac{x-\zeta}{\theta} \right)^{-\frac{1}{\xi}-1} & \xi \neq 0 \\ \frac{1}{|\theta|} \exp\left(-\frac{x-\zeta}{\theta} \right) & \xi = 0 \end{cases}$$

$$= \text{PowerFn}(x \; ; \zeta - \tfrac{\theta}{\xi}, \tfrac{\theta}{\xi}, -\tfrac{1}{\xi})$$

q-exponential (generalized Pareto) distribution is an alternative parameterization of the power function distribution, utilizing the Tsallis generalized q-exponential function, $\exp_q(x)$ (§D).

$$QExp(x \; ; \; \zeta, \theta, q) \tag{5.3}$$

$$= \frac{(2-q)}{|\theta|} \exp_q\left(-\frac{x-\zeta}{\theta}\right)$$

$$= \begin{cases} \frac{(2-q)}{|\theta|}\left(1 - (1-q)\frac{x-\zeta}{\theta}\right)^{\frac{1}{1-q}} & q \neq 1 \\ \frac{1}{|\theta|}\exp\left(-\frac{x-\zeta}{\theta}\right) & q = 1 \end{cases}$$

$$= PowerFn(x \; ; \; \zeta + \frac{\theta}{1-q}, -\frac{\theta}{1-q}, \frac{2-q}{1-q})$$

for x, ζ, θ, q in $\mathbb{R}$

Special cases: Positive β

Pearson [7, 2] noted two special cases, the monotonically decreasing **Pearson type VIII** $0 < \beta < 1$, and the monotonically increasing **Pearson type IX** distribution [7, 2] with $\beta > 1$.

Wedge distribution [2]:

$$Wedge(x \; ; \; a, s) = 2 \; \text{sgn}(s)\frac{x-a}{s^2} \tag{5.4}$$

$$= PowerFn(x \; ; \; a, s, 2)$$

With a positive scale we obtain an **ascending wedge** (right triangular) distribution, and with negative scale a **descending wedge** (left triangular).

Special cases: Negative β

Pareto (Pearson XI, Pareto type I) distribution [18, 7, 2]:

$$Pareto(x \; ; \; a, s, \gamma) = \left|\frac{\bar{\beta}}{s}\right|\left(\frac{x-a}{s}\right)^{-\bar{\beta}-1} \qquad \bar{\beta} > 0 \tag{5.5}$$

$$x > a + s, \; s > 0$$

$$x < a + s, \; s < 0$$

$$= PowerFn(x \; ; \; a, s, -\bar{\beta})$$

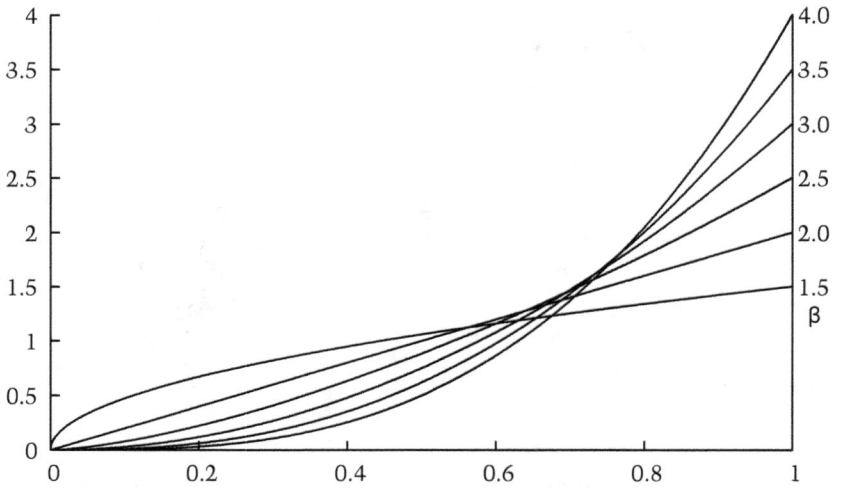

Figure 10: Pearson type IX, PowerFn(x ; $0, 1, \beta$), $\beta > 1$

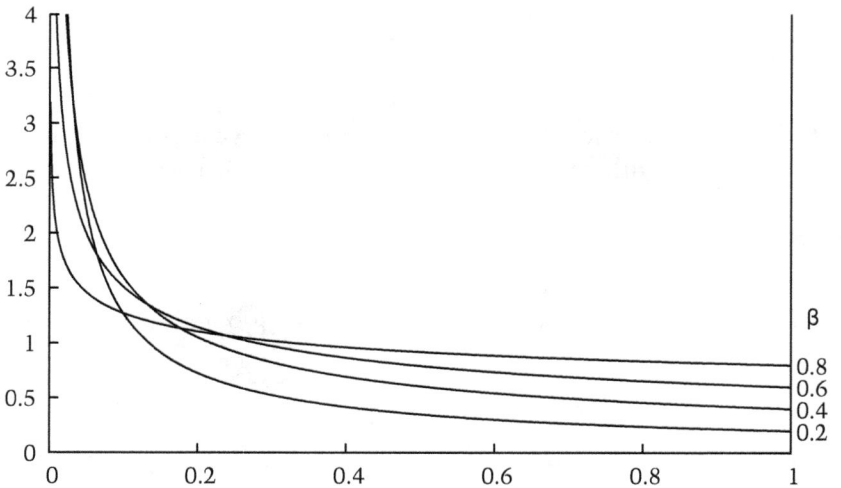

Figure 11: Pearson type VIII, PowerFn(x ; $0, 1, \beta$), $0 < \beta < 1$.

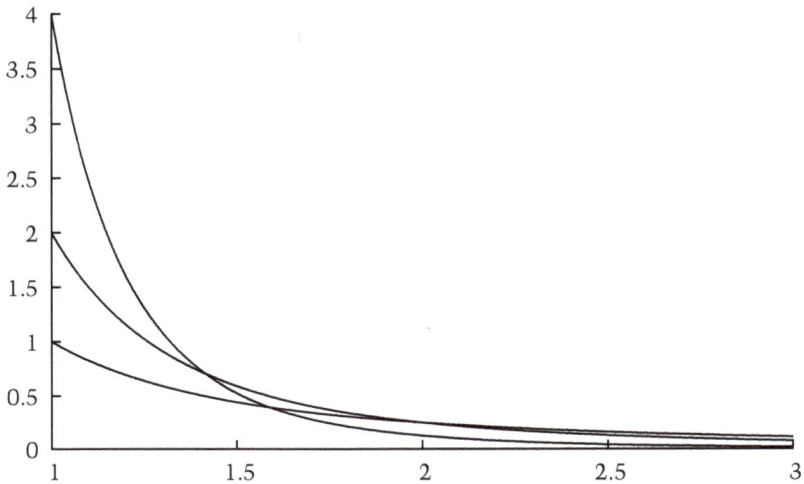

Figure 12: Pareto distributions, $\text{Pareto}(x\ ;\ 0, 1, \bar{\beta})$, $\bar{\beta}$ left axis.

The most important special case is the Pareto distribution, which has a semi-infinite support with a power-law tail. The Zipf distribution is the discrete analog of the Pareto distribution.

Lomax (Pareto type II, ballasted Pareto) distribution [19]:

$$\text{Lomax}(x\ ;\ a, s, \bar{\beta}) = \frac{\bar{\beta}}{|s|}\left(1 + \frac{x - a}{s}\right)^{-\bar{\beta} - 1} \tag{5.6}$$
$$= \text{Pareto}(x\ ;\ a - s, s, \bar{\beta})$$
$$= \text{PowerFn}(x\ ;\ a - s, s, -\bar{\beta})$$

Originally explored as a model of business failure. The alternative name "ballasted Pareto" arises since this distribution is a shifted Pareto distribution (5.5) whose origin is fixed at zero, and no longer moves with changes in scale.

Table 5.1: Special cases of the power function distribution

		a	s	β
(5.1)	power function			
(5.5)	Pareto	.	.	<0
(5.8)	uniform prime	.	.	-1
(5.1)	Pearson type VIII	0	.	$(0, 1)$
(1.1)	uniform	.	.	1
(5.1)	Pearson type IX	0	.	>1
(5.4)	wedge	.	.	2
(2.1)	exponential	.	.	$+\infty$

Exponential ratio distribution [1]:

$$\text{ExpRatio}(x \; ; \; s) = \frac{1}{|s|} \frac{1}{\left(1 + \frac{x}{s}\right)^2} \tag{5.7}$$

$$= \text{Lomax}(x \; ; \; 0, s, 1)$$

$$= \text{PowerFn}(x \; ; \; -s, s, 1)$$

Arises as the ratio of independent exponential distributions (p 29).

Uniform-prime distribution [20, 1]:

$$\text{UniPrime}(x \; ; \; a, s) = \frac{1}{|s|} \frac{1}{\left(1 + \frac{x-a}{s}\right)^2} \tag{5.8}$$

$$= \text{Lomax}(x \; ; \; a, s, 1)$$

$$= \text{PowerFn}(x \; ; \; a - s, s, -1)$$

An exponential ratio (5.7) distribution with a shift parameter. So named since this distribution is related to the uniform distribution as beta is to beta prime. The ordering distribution (§C) of the beta-prime distribution.

Limits and subfamilies

With $\beta = 1$ we recover the uniform distribution.

$$\text{PowerFn}(a, s, 1) \sim \text{Uniform}(a, s)$$

As β limits to infinity, the power function distribution limits to the exponential distribution (2.1).

$$\text{Exp}(x \; ; \; \nu, \lambda) = \lim_{\beta \to \infty} \text{PowerFn}(x \; ; \; \nu - \beta\lambda, \beta\lambda, \beta)$$

$$= \lim_{\beta \to \infty} \left| \frac{1}{\lambda} \right| \left(1 + \frac{x - \nu}{\beta\lambda} \right)^{\beta - 1}$$

Recall that $\lim_{c \to \infty} \left(1 + \frac{x}{c} \right)^c = e^x$.

Interrelations

With positive β, the power function distribution is a special case of the beta distribution (12.1), with negative beta, a special case of the beta prime distribution (13.1), and with either sign a special case of the generalized beta (17.1) and unit gamma (10.1) distributions.

$$\text{PowerFn}(x \; ; \; a, s, \beta)$$
$$= \text{GenBeta}(x \; ; \; a, s, 1, 1, \beta)$$
$$= \text{GenBeta}(x \; ; \; a, s, \beta, 1, 1) \qquad\qquad \beta > 0$$
$$= \text{Beta}(x \; ; \; a, s, \beta, 1) \qquad\qquad \beta > 0$$
$$= \text{GenBeta}(x \; ; \; a + s, s, 1, -\beta, -1) \qquad\qquad \beta < 0$$
$$= \text{BetaPrime}(x \; ; \; a + s, s, 1, -\beta) \qquad\qquad \beta < 0$$
$$= \text{UnitGamma}(x \; ; \; a, s, 1, \beta)$$

The order statistics (§C) of the power function distribution yields the generalized beta distribution (17.1).

$$\text{OrderStatistic}_{\text{PowerFn}(a,s,\beta)}(x \; ; \; \alpha, \gamma) = \text{GenBeta}(x \; ; \; a, s, \alpha, \gamma, \beta)$$

Since the power function distribution is a special case of the generalized beta distribution (17.1),

$$\text{GenBeta}(x \; ; \; a, s, \alpha, 1, \beta) = \text{PowerFn}(x \; ; \; a, s, \alpha\beta)$$

it follows that the power function family is closed under maximization for $\frac{\beta}{s} > 0$ and minimization for $\frac{\beta}{s} < 0$.

The product of independent power function distributions (With zero location parameter, and the same β) is a unit-gamma distribution (10.1) [21].

$$\prod_{i=1}^{\alpha} \text{PowerFn}_i(0, s_i, \beta) \sim \text{UnitGamma}(0, \prod_{i=1}^{\alpha} s_i, \alpha, \beta)$$

Consequently, the geometric mean of independent, anchored power function distributions (with common β) is also unit-gamma.

$$\sqrt[\alpha]{\prod_{i=1}^{\alpha} \text{PowerFn}_i(0, s_i, \beta)} \sim \text{UnitGamma}(0, \prod_{i=1}^{\alpha} s_i, \alpha, \alpha\beta)$$

The power function distribution can be obtained from the Weibull transform $x \rightarrow (\frac{x-a}{s})^\beta$ of the uniform distribution (1.1).

$$\text{PowerFn}(a, s, \beta) \sim a + s \; \text{StdUniform}()^{\frac{1}{\beta}}$$

The power function distribution limits to the exponential distribution (§2).

$$\text{Exp}(x \; ; a, \theta) = \lim_{\beta \rightarrow \infty} \text{PowerFn}(x \; ; a + \beta\theta, -\beta\theta, \beta)$$

Table 5.2: Properties of the power function distribution

Properties

notation	$\text{PowerFn}(x \; ; a, s, \beta)$	
PDF	$\left\lvert\dfrac{\beta}{s}\right\rvert \left(\dfrac{x-a}{s}\right)^{\beta-1}$	
CDF/CCDF	$\left(\dfrac{x-a}{s}\right)^{\beta}$	$\frac{s}{\beta} > 0 \;/\; \frac{s}{\beta} < 0$
parameters	a, s, β in $\mathbb{R}$	
support	$x \in [a, a+s]$	$s > 0, \; \beta > 0$
	$x \in [a+s, a]$	$s < 0, \; \beta > 0$
	$x \in [a+s, +\infty]$	$s > 0, \; \beta < 0$
	$x \in [-\infty, a+s]$	$s < 0, \; \beta < 0$
mode	a	$\beta > 0$
	$a + s$	$\beta < 0$
mean	$a + \dfrac{s\beta}{\beta+1}$	$\beta \notin [-1, 0]$
variance	$\dfrac{s^2\beta}{(\beta+1)^2(\beta+2)}$	$\beta \notin [-2, 0]$
skew	$\text{sgn}(\tfrac{\beta}{s}) \dfrac{2(1-\beta)}{(\beta+3)} \sqrt{\dfrac{\beta+2}{\beta}}$	$\beta \notin [-3, 0]$
ex. kurtosis	$\dfrac{6(\beta^3 - \beta^2 - 6\beta + 2)}{\beta(\beta+3)(\beta+4)}$	$\beta \notin [-4, 0]$
MGF	undefined	

6 Log-Normal Distribution

Log-normal (Galton, Galton-McAlister, anti-log-normal, Cobb-Douglas, log-Gaussian, logarithmic-normal, logarithmico-normal, Λ, Gibrat) distribution [22, 23, 2] is a three parameter, continuous, univariate, unimodal probability density with semi-infinite support. The functional form in the standard parameterization is

$$\text{LogNormal}(x \; ; a, \vartheta, \beta) \tag{6.1}$$
$$= \frac{|\beta|}{\sqrt{2\pi\vartheta^2}} \left(\frac{x-a}{\vartheta}\right)^{-1} \exp\left\{-\frac{1}{2}\left(\beta \ln \frac{x-a}{\vartheta}\right)^2\right\}$$
$$\text{for } x, a, \vartheta, \beta \text{ in } \mathbb{R},$$
$$\tfrac{x-a}{\vartheta} > 0$$

The log-normal is so called because the log transform of the log-normal variate is a normal random variable. The distribution should, perhaps, be more accurately called the anti-log-normal distribution, but the nomenclature is now standard.

Special cases

The **anchored log-normal** (two-parameter log-normal) distribution ($a = 0$) arises from the multiplicative version of the central limit theorem: When the sum of independent random variables limits to normal, the product of those random variables limits to log-normal. With $a = 0$, $\vartheta = 1$, $\sigma = 1$ we obtain the **standard log-normal** (Gibrat) distribution [24].

Interrelations

The log-normal forms a location-scale-power distribution family.

$$\text{LogNormal}(a, \vartheta, \beta) \sim a + \vartheta \, \text{StdLogNormal}()^{\frac{1}{\beta}}$$

The log-normal distribution is the anti-log transform of a normal random variable.

$$\text{LogNormal}(a, \vartheta, \beta) \sim a + \exp\left(-\text{Normal}(-\ln \vartheta, \tfrac{1}{\beta})\right)$$

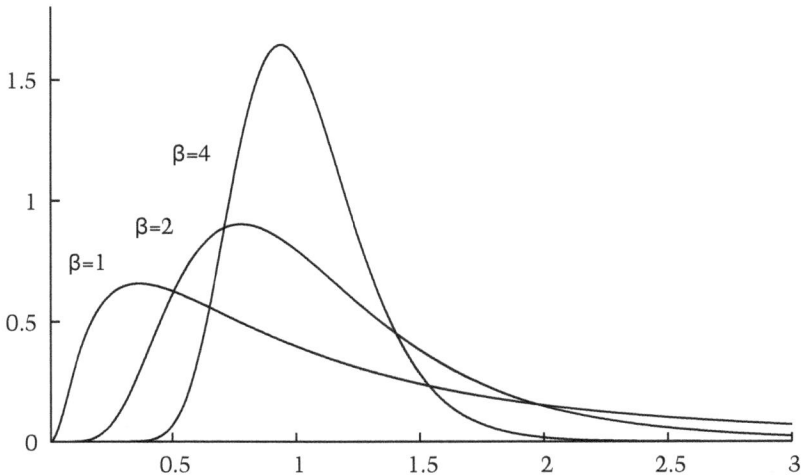

Figure 13: Log normal distributions, LogNormal$(x \; ; 0, 1, \beta)$

Because of this close connection to the normal distribution, the log-normal is often parameterized with the mean and standard deviation of the corresponding normal distribution, $\mu = \ln \vartheta$, $\sigma = 1/\beta$ rather than standard scale and power parameters.

The log-normal distribution is a limiting form of the Unit gamma (10.1) and Amoroso (11.1), distributions (And therefore also of the generalized beta and generalized beta prime distributions) and limits to the normal distribution (§D).

$$\text{Normal}(x \; ; \mu, \sigma) = \lim_{\beta \to \infty} \text{LogNormal}(x \; ; \mu + \beta\sigma, -\beta\sigma, \beta)$$

A product of log-normal distributions (With zero location parameter) is again a log-normal distribution. This follows from the fact that the sum of normal distributions is normal.

$$\prod_{i=1}^{n} \text{LogNormal}_i(0, \vartheta_i, \beta_i) \sim \text{LogNormal}_i(0, \prod_{i=1}^{n} \vartheta_i, (\sum_{i=0}^{n} \beta_i^{-2})^{-\frac{1}{2}})$$

Table 6.1: Properties of the log-normal distribution

Properties	
notation	$\text{LogNormal}(x \, ; \, a, \vartheta, \beta)$
PDF	$\dfrac{\lvert\beta\rvert}{\sqrt{2\pi\vartheta^2}} \left(\dfrac{x-a}{\vartheta}\right)^{-1} \exp\left\{-\dfrac{1}{2}\left(\beta\ln\dfrac{x-a}{\vartheta}\right)^2\right\}$
CDF/CCDF	$\dfrac{1}{2} + \dfrac{1}{2}\text{erf}\left(\dfrac{1}{\sqrt{2}}\beta\ln\dfrac{x-a}{\vartheta}\right)$ $\qquad\qquad \vartheta > 0 \, / \, \vartheta < 0$
parameters	$a, \, \vartheta, \, \beta \text{ in } \mathbb{R}$
support	$x \in [a, +\infty] \quad \vartheta > 0$
	$x \in [-\infty, a] \quad \vartheta < 0$
median	$a + \vartheta$
mode	$a + \vartheta e^{-\beta^{-2}}$
mean	$a + \vartheta e^{\frac{1}{2}\beta^{-2}}$
variance	$\vartheta^2 (e^{\beta^{-2}} - 1) e^{\beta^{-2}}$
skew	$\text{sgn}(\vartheta)\, (e^{\beta^{-2}} + 2)\sqrt{e^{\beta^{-2}} - 1}$
ex. kurtosis	$e^{4\beta^{-2}} + 2e^{3\beta^{-2}} + 3e^{2\beta^{-2}} - 6$
entropy	$\dfrac{1}{2} + \dfrac{1}{2}\ln(2\pi\beta^{-2}) + \ln\lvert\vartheta\rvert$
MGF	doesn't exist in general
CF	no simple closed form expression

7   GAMMA DISTRIBUTION

Gamma (Γ, Pearson type III) distribution [4, 5, 2] :

$$\text{Gamma}(x\ ;\ a, \theta, \alpha) = \frac{1}{\Gamma(\alpha)|\theta|} \left(\frac{x - a}{\theta} \right)^{\alpha - 1} \exp\left\{ -\frac{x - a}{\theta} \right\} \qquad (7.1)$$

$$\text{for } x,\ a, \theta, \alpha \text{ in } \mathbb{R}, \quad \alpha > 0$$

$$= \text{Amoroso}(x\ ;\ a, \theta, \alpha, 1)$$

The name of this distribution derives from the normalization constant.

Special cases

Special cases of the beta prime distribution are listed in table 11.1, under $\beta = 1$.

The gamma distribution often appear as a solution to problems in statistical physics. For example, the energy density of a classical ideal gas, or the **Wien** (Vienna) distribution $\text{Wien}(x\ ;\ T) = \text{Gamma}(x\ ;\ 0, T, 4)$, an approximation to the relative intensity of black body radiation as a function of the frequency. The **Erlang** (m-Erlang) distribution [25] is a gamma distribution with integer α, which models the waiting time to observe α events from a Poisson process with rate $1/\theta$ ($\theta > 0$). For $\alpha = 1$ we obtain an exponential distribution (2.1).

Standard gamma (standard Amoroso) distribution [2]:

$$\text{StdGamma}(x\ ;\ \alpha) = \frac{1}{\Gamma(\alpha)} x^{\alpha - 1} e^{-x} \qquad (7.2)$$

$$= \text{Gamma}(x\ ;\ 0, 1, \alpha)$$

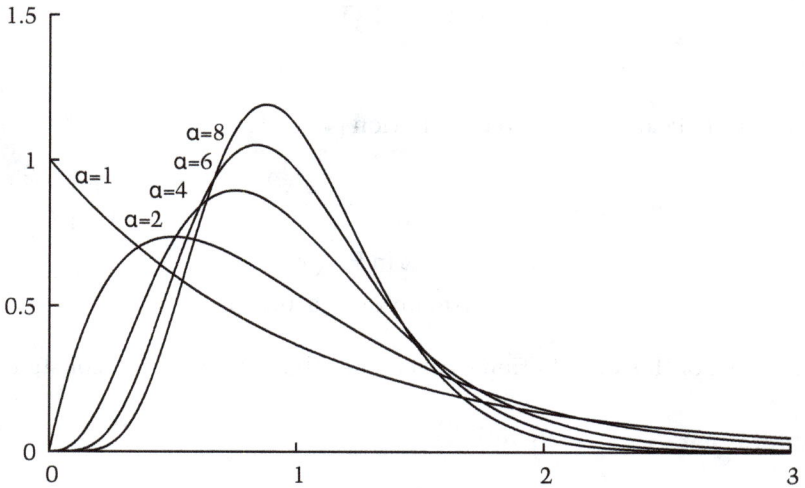

Figure 14: Gamma distributions, unit variance Gamma$(x ; \frac{1}{\alpha}, \alpha)$

Chi-square (χ^2) distribution [26, 2]:

$$\text{ChiSqr}(x ; k) = \frac{1}{2\Gamma(\frac{k}{2})}\left(\frac{x}{2}\right)^{\frac{k}{2}-1} \exp\left\{-\left(\frac{x}{2}\right)\right\} \qquad (7.3)$$

$$\text{for positive integer } k$$

$$= \text{Gamma}(x ; 0, 2, \tfrac{k}{2})$$

$$= \text{Stacy}(x ; 2, \tfrac{k}{2}, 1)$$

$$= \text{Amoroso}(x ; 0, 2, \tfrac{k}{2}, 1)$$

The distribution of a sum of squares of k independent standard normal random variables. The chi-square distribution is important for statistical hypothesis testing in the frequentist approach to statistical inference.

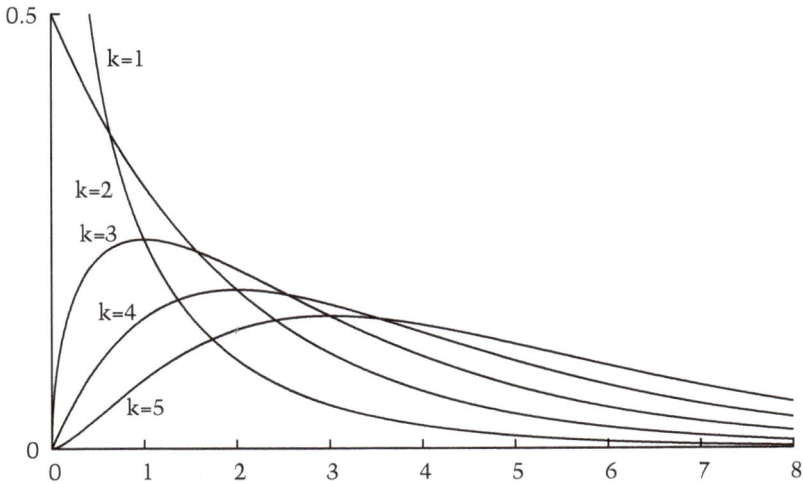

Figure 15: Chi-square distributions, ChiSqr(x ; k)

Scaled chi-square distribution [27]:

$$\text{ScaledChiSqr}(x\ ;\ \sigma, k) = \frac{1}{2\sigma^2\Gamma(\frac{k}{2})}\left(\frac{x}{2\sigma^2}\right)^{\frac{k}{2}-1}\exp\left\{-\left(\frac{x}{2\sigma^2}\right)\right\} \qquad (7.4)$$

for positive integer k

$$= \text{Stacy}(x\ ;\ 2\sigma^2, \tfrac{k}{2}, 1)$$
$$= \text{Gamma}(x\ ;\ 0, 2\sigma^2, \tfrac{k}{2})$$
$$= \text{Amoroso}(x\ ;\ 0, 2\sigma^2, \tfrac{k}{2}, 1)$$

The distribution of a sum of squares of k independent normal random variables with variance σ^2.

Table 7.1: Special cases of the gamma family

(7.1)	gamma	a	θ	α
(7.1)	Erlang	0	>0	n
(7.2)	standard gamma	0	1	.
(7.5)	Porter-Thomas	0	2	$\frac{1}{2}$
(7.4)	scaled chi-square	0	.	$\frac{1}{2}k$
(7.3)	chi-square	0	2	$\frac{1}{2}k$
(2.1)	exponential	.	.	1
(7.1)	Wien	0	.	4

(k, n positive integers)

Porter-Thomas distribution [28]:

$$PorterThomas(x \; ; \sigma) = \frac{1}{2\sigma^2\Gamma(\frac{1}{2})}\left(\frac{x}{2\sigma^2}\right)^{-\frac{1}{2}} \exp\left\{-\left(\frac{x}{2\sigma^2}\right)\right\} \tag{7.5}$$

$$= Stacy(x \; ; 2\sigma^2, \tfrac{1}{2}, 1)$$
$$= Gamma(x \; ; 0, 2\sigma^2, \tfrac{1}{2})$$
$$= Amoroso(x \; ; 0, 2\sigma^2, \tfrac{1}{2}, 1)$$

A chi-square distribution with a single degree of freedom. Used to model fluctuations in decay mode strengths of excited nuclei [28]

Interrelations

Gamma distributions with common scale obey an addition property:

$$Gamma_1(0, \theta, \alpha_1) + Gamma_2(0, \theta, \alpha_2) \sim Gamma_3(0, \theta, \alpha_1 + \alpha_2)$$

The sum of two independent, gamma distributed random variables (with common θ's, but possibly different α's) is again a gamma random variable [2].

The Amoroso distribution can be obtained from the standard gamma

Table 7.2: Properties of the gamma distribution

Properties		
notation	$\mathrm{Gamma}(x\,;\,a,\theta,\alpha)$	
PDF	$\dfrac{1}{\Gamma(\alpha)|\theta|}\left(\dfrac{x-a}{\theta}\right)^{\alpha-1}\exp\left\{-\dfrac{x-a}{\theta}\right\}$	
CDF / CCDF	$1-Q\left(\alpha,\frac{x-a}{\theta}\right)$	$\theta>0\,/\,\theta<0$
parameters	$a,\ \theta,\ \alpha,\ \mathrm{in}\ \mathbb{R},\ \alpha>0$	
support	$x\geqslant a$	$\theta>0$
	$x\leqslant a$	$\theta<0$
mode	$a+\theta(\alpha-1)$	$\alpha\geqslant 1$
	a	$\alpha\leqslant 1$
mean	$a+\theta\alpha$	
variance	$\theta^2\alpha$	
skew	$\mathrm{sgn}(\theta)\,\dfrac{2}{\sqrt{\alpha}}$	
ex. kurtosis	$\dfrac{6}{\alpha}$	
entropy	$\ln\left(|\theta|\Gamma(\alpha)\right)+\alpha+(1-\alpha)\psi(\alpha)$	
MGF	$e^{at}(1-\theta t)^{-\alpha}$	
CF	$e^{iat}(1-i\theta t)^{-\alpha}$	

distribution by the Weibull change of variables, $x \rightarrow \left(\frac{x-a}{\theta}\right)^{\beta}$.

$$\text{Amoroso}(a, \theta, \alpha, \beta) \sim a + \theta \left[\text{StdGamma}(\alpha)\right]^{1/\beta}$$

For large α the gamma distribution limits to normal (4.1).

$$\text{Normal}(x \; ; \mu, \sigma) = \lim_{\alpha \to \infty} \text{Gamma}(x \; ; \mu - \sigma\sqrt{\alpha}, \frac{\sigma}{\sqrt{\alpha}}, \alpha)$$

Conversely, the sum of squares of normal distributions is a gamma distribution. See chi-square (7.3).

$$\sum_{i=1,k} \text{StdNormal}_i()^2 \sim \text{ChiSqr}(k) \sim \text{Gamma}(0, 2, \frac{k}{2})$$

A large variety of distributions can be obtained from transformations of 1 or 2 gamma distributions, which is convenient for generating pseudorandom numbers from those distributions (See appendix (§E)).

$$\text{Normal}(\mu, \sigma) \sim \mu + \sigma \ \text{Sgn}() \ \sqrt{2 \, \text{StdGamma}(\tfrac{1}{2})} \qquad (4.1)$$

$$\text{GammaExp}(a, s, \alpha) \sim a - s \ln\big(\text{StdGamma}(\alpha)\big) \qquad (8.1)$$

$$\text{PearsonVII}(a, s, m) \sim a + s \ \text{Sgn}() \sqrt{\frac{\text{StdGamma}_1(\tfrac{1}{2})}{\text{StdGamma}_2(m - \tfrac{1}{2})}} \qquad (9.1)$$

$$\text{Cauchy}(a, s) \sim a + s \ \text{Sgn}() \sqrt{\frac{\text{StdGamma}_1(\tfrac{1}{2})}{\text{StdGamma}_2(\tfrac{1}{2})}} \qquad (9.6)$$

$$\text{UnitGamma}(a, s, \alpha, \beta) \sim a + s \ \exp\big(-\tfrac{1}{\beta} \text{StdGamma}(\alpha)\big) \qquad (10.1)$$

$$\text{Beta}(a, s, \alpha, \gamma) \sim a + s \left(1 + \frac{\text{StdGamma}_2(\gamma)}{\text{StdGamma}_1(\alpha)}\right)^{-1} \qquad (12.1)$$

$$\text{BetaPrime}(a, s, \alpha, \gamma) \sim a + s \ \frac{\text{StdGamma}_1(\alpha)}{\text{StdGamma}_2(\gamma)} \qquad (13.1)$$

$$\text{Amoroso}(a, \theta, \alpha, \beta) \sim a + \theta \ \text{StdGamma}(\alpha)^{\frac{1}{\beta}} \qquad (11.1)$$

$$\text{BetaExp}(a, s, \alpha, \gamma) \sim a - s \ \ln\left(1 + \frac{\text{StdGamma}_2(\gamma)}{\text{StdGamma}_1(\alpha)}\right)^{-1} \qquad (14.1)$$

$$\text{BetaLogistic}(a, s, \alpha, \gamma) \sim a - s \ln\left(\frac{\text{StdGamma}_1(\alpha)}{\text{StdGamma}_2(\gamma)}\right) \qquad (15.1)$$

$$\text{GenBeta}(a, s, \alpha, \gamma, \beta) \sim a + s \left(1 + \frac{\text{StdGamma}_2(\gamma)}{\text{StdGamma}_1(\alpha)}\right)^{-\frac{1}{\beta}} \qquad (17.1)$$

$$\text{GenBetaPrime}(a, s, \alpha, \gamma, \beta) \sim a + s \left(\frac{\text{StdGamma}_1(\alpha)}{\text{StdGamma}_2(\gamma)}\right)^{\frac{1}{\beta}} \qquad (18.1)$$

Here, Sgn() is the sign (or Rademacher) discrete random variable: 50% chance -1, 50% chance $+1$.

8   Gamma-Exponential Distribution

The **gamma-exponential** (log-gamma, generalized Gompertz, generalized Gompertz-Verhulst type I, Coale-McNeil, exponential gamma) distribution [29, 30, 3, 31] is a three parameter, continuous, univariate, unimodal probability density with infinite support. The functional form in the most straightforward parameterization is

$$
\text{GammaExp}(x \; ; \nu, \lambda, \alpha) \tag{8.1}
$$
$$
= \frac{1}{\Gamma(\alpha)|\lambda|} \exp\left\{-\alpha\left(\frac{x-\nu}{\lambda}\right) - \exp\left(-\frac{x-\nu}{\lambda}\right)\right\}
$$
$$
\text{for } x, \; \nu, \; \lambda, \; \alpha, \text{ in } \mathbb{R}, \; \alpha > 0,
$$
$$
\text{support } -\infty \leqslant x \leqslant \infty
$$

The three real parameters consist of a location parameter ν, a scale parameter λ, and a shape parameter α.

Note that this distribution is often called the "log-gamma" distribution. This naming convention is the opposite of that used for the log-normal distribution (6.1). The name "log-gamma" has also been used for the anti-log transform of the generalized gamma distribution, which leads to the unit-gamma distribution (10.1).

Also note that the gamma-exponential is often defined with the sign of the scale λ flipped. The parameterization used here is consistent with other log-transformed distributions. (See Log and anti-log transformation, p.169)

Special cases

Standard gamma-exponential distribution:

$$
\text{StdGammaExp}(x \; ; \alpha) = \frac{1}{\Gamma(\alpha)} \exp\{-\alpha x - \exp(-x)\} \tag{8.2}
$$
$$
= \text{GammaExp}(x \; ; 0, 1, \alpha)
$$

The gamma-exponential distribution with zero location and unit scale.

Table 8.1: Special cases of the gamma-exponential family

(8.1)	gamma-exponential	ν	λ	α
(8.2)	standard gamma-exponential	0	1	α
(8.3)	chi-square-exponential	$\ln 2$	1	$\frac{k}{2}$
(8.4)	generalized Gumbel	.	.	n
(8.5)	Gumbel	.	.	1
(8.6)	standard Gumbel	0	1	1
(8.7)	BHP	.	.	$\frac{\pi}{2}$
(8.8)	Moyal	.	.	$\frac{1}{2}$

Chi-square-exponential (log-chi-square) distribution [27]:

$$\text{ChiSqrExp}(x\; ; k) = \frac{1}{2^{\frac{k}{2}}\Gamma(\frac{k}{2})}\exp\left\{-\frac{k}{2}x - \frac{1}{2}\exp(-x)\right\}$$

$$\text{for positive integer } k \qquad (8.3)$$

$$= \text{GammaExp}(x\; ; \ln 2, 1, \tfrac{k}{2})$$

The log transform of the chi-square distribution (7.3).

Generalized Gumbel distribution [32, 3]:

$$\text{GenGumbel}(x\; ; u, \lambda, n) \qquad (8.4)$$

$$= \frac{n^n}{\Gamma(n)|\lambda|}\exp\left\{-n\left(\frac{x-u}{\lambda}\right) - n\exp\left(-\frac{x-u}{\lambda}\right)\right\}$$

$$\text{for positive integer } n$$

$$= \text{GammaExp}(x\; ; u - \lambda \ln n, \lambda, n)$$

The limiting distribution of the nth largest value of a large number of unbounded identically distributed random variables whose probability distribution has an exponentially decaying tail.

Gumbel (Fisher-Tippett type I, Fisher-Tippett-Gumbel, Gumbel-Fisher-Tippett, FTG, log-Weibull, extreme value (type I), doubly exponential, dou-

Table 8.2: Properties of the gamma-exponential distribution

Properties		
notation	$\mathrm{GammaExp}(x\,;\,\nu, \lambda, \alpha)$	
PDF	$\dfrac{1}{\Gamma(\alpha)|\lambda|} \exp\left\{ -\alpha\left(\dfrac{x-\nu}{\lambda}\right) - \exp\left(-\dfrac{x-\nu}{\lambda}\right)\right\}$	
CDF/CCDF	$Q\left(\alpha, e^{-\frac{x-\nu}{\lambda}}\right)$	$\lambda > 0 \,/\, \lambda < 0$
parameters	$\nu, \lambda, \alpha, \text{ in } \mathbb{R}, \alpha > 0,$	
support	$x \in [-\infty, +\infty]$	
mode	$\nu - \lambda \ln \alpha$	
mean	$\nu - \lambda\psi(\alpha)$	
variance	$\lambda^2 \psi_1(\alpha)$	
skew	$-\operatorname{sgn}(\lambda)\dfrac{\psi_2(\alpha)}{\psi_1(\alpha)^{3/2}}$	
ex. kurtosis	$\dfrac{\psi_3(\alpha)}{\psi_1(\alpha)^2}$	
MGF	$e^{\nu t}\dfrac{\Gamma(\alpha - \lambda t)}{\Gamma(\alpha)}$	[3]
CF	$e^{i\nu t}\dfrac{\Gamma(\alpha - i\lambda t)}{\Gamma(\alpha)}$	

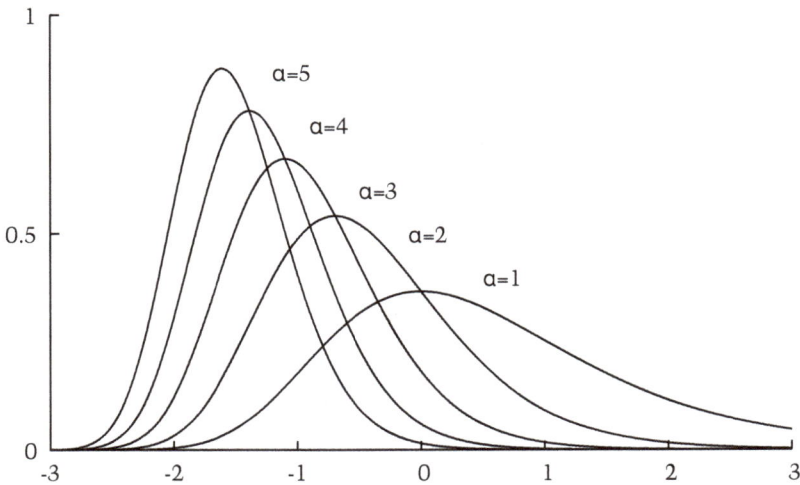

Figure 16: Gamma exponential distributions, GammaExp$(x ; 0, 1, \alpha)$

ble exponential) distribution [33, 32, 3]:

$$\text{Gumbel}(x ; u, \lambda) = \frac{1}{|\lambda|} \exp\left\{-\left(\frac{x-u}{\lambda}\right) - \exp\left(-\frac{x-u}{\lambda}\right)\right\} \quad (8.5)$$
$$= \text{GammaExp}(x ; u, \lambda, 1)$$

This is the asymptotic extreme value distribution for variables of "exponential type", unbounded with finite moments [32]. With positive scale $\lambda > 0$, this is an extreme value distribution of the maximum, with negative scale $\lambda < 0$ an extreme value distribution of the minimum. Note that the Gumbel is sometimes defined with the negative of the scale used here.

The term "double exponential distribution" can refer to either Laplace or Gumbel distributions [3].

Standard Gumbel (Gumbel) distribution [32]:

$$\text{StdGumbel}(x) = \exp\left\{-x - e^{-x}\right\} \quad (8.6)$$
$$= \text{GammaExp}(x ; 0, 1, 1)$$

The Gumbel distribution with zero location and a unit scale.

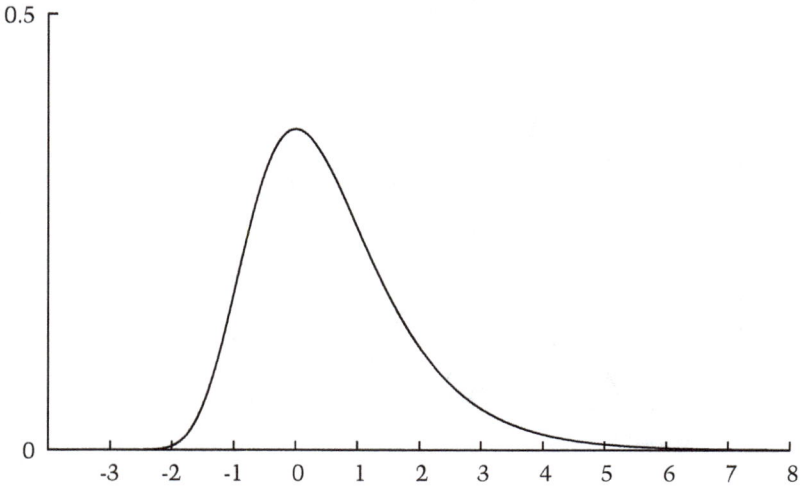

Figure 17: Standard Gumbel distribution, StdGumbel(x)

BHP (Bramwell-Holdsworth-Pinton) distribution [34, 35]:

$$\mathrm{BHP}(x\ ;\ \nu,\lambda) = \frac{1}{\Gamma(\frac{\pi}{2})|\lambda|}\exp\left\{-\frac{\pi}{2}\left(\frac{x-\nu}{\lambda}\right) - \exp\left(-\frac{x-\nu}{\lambda}\right)\right\}$$
$$= \mathrm{GammaExp}(x\ ;\ \nu,\lambda,\frac{\pi}{2}) \tag{8.7}$$

Proposed as a model of rare fluctuations in turbulence and other correlated systems.

Moyal distribution [36]:

$$\mathrm{Moyal}(x\ ;\ \mu,\lambda) = \frac{1}{\sqrt{2\pi}|\lambda|}\exp\left\{-\tfrac{1}{2}\left(\frac{x-\mu}{\lambda}\right) - \tfrac{1}{2}\exp\left(-\frac{x-\mu}{\lambda}\right)\right\} \tag{8.8}$$
$$= \mathrm{GammaExp}(x\ ;\ \mu+\lambda\ln 2,\lambda,\tfrac{1}{2})$$

Introduced as analytic approximation to the Landau distribution (21.11) [36].

Interrelations

The name "log-gamma" arises because the standard log-gamma distribution is the logarithmic transform of the standard gamma distribution

$$\text{StdGammaExp}(\alpha) \sim -\ln\Big(\text{StdGamma}(\alpha)\Big)$$
$$\text{GammaExp}(\nu, \lambda, \alpha) \sim -\ln\Big(\text{Amoroso}(0, e^{-\nu}, \alpha, \tfrac{1}{\lambda})\Big)$$

The difference of two gamma-exponential distribution (with common scale) is a beta-logistic distribution (15.1) [3].

$$\text{BetaLogistic}(x \; ; \; \zeta_1 - \zeta_2, \lambda, \alpha, \gamma) \sim \text{GammaExp}_1(x \; ; \; \zeta_1, \lambda, \alpha)$$
$$- \text{GammaExp}_2(x \; ; \; \zeta_2, \lambda, \gamma)$$

It follows that the difference of two Gumbel distributions (8.5) is a logistic distribution (15.5).

The gamma-exponential distribution is a limit of the Amoroso distribution (11.1), and itself contains the normal (4.1) distribution as a limiting case.

$$\lim_{\alpha \to \infty} \text{GammaExp}(x \; ; \; \mu + \sigma\sqrt{\alpha}\ln\alpha, \sigma\sqrt{\alpha}, \alpha) = \text{Normal}(x \; ; \; \mu, \sigma)$$

9   Pearson VII Distribution

The **Pearson type VII** distribution [7] is a three parameter, continuous, univariate, unimodal, symmetric probability distribution, with infinite support. The functional form in the most straight forward parameterization is

$$\mathrm{PearsonVII}(x \; ; \; a, s, m) = \frac{1}{|s|B(m - \frac{1}{2}, \frac{1}{2})} \left(1 + \left(\frac{x - a}{s} \right)^2 \right)^{-m} \tag{9.1}$$

$$m > \tfrac{1}{2}$$

$$= \mathrm{PearsonIV}(x \; ; \; a, s, m, 0)$$

This distribution family is notable for having long power-law tails in both directions.

Special cases

Student's t (Student, t, Student-Fisher, Fisher) distribution [37, 38, 39, 40] :

$$\mathrm{StudentsT}(x \; ; \; k) = \frac{1}{\sqrt{k}B(\frac{1}{2}, \frac{1}{2}k)} \left(1 + \frac{x^2}{k} \right)^{-\frac{1}{2}(k+1)} \tag{9.2}$$

$$= \mathrm{PearsonVII}(x \; ; \; 0, \sqrt{k}, \tfrac{1}{2}(k+1))$$

$$\text{integer } k \geqslant 0$$

The distribution of the statistic t, which arises when considering the error of samples means drawn from normal random variables.

$$t = \sqrt{n} \frac{\bar{x} - \mu}{\bar{s}}$$

$$\bar{x} = \tfrac{1}{n} \sum_{i=1}^{n} \mathrm{Normal}_i(\mu, \sigma)$$

$$\bar{s}^2 = \tfrac{1}{n-1} \sum_{i=1}^{n} \left(\mathrm{Normal}_i(\mu, \sigma) - \bar{x} \right)^2$$

Here, $\bar{x}$ is the sample mean of n independent normal (4.1) random variables with mean μ and variance σ^2, $\bar{s}$ is the sample variance, and $k = n - 1$ is the

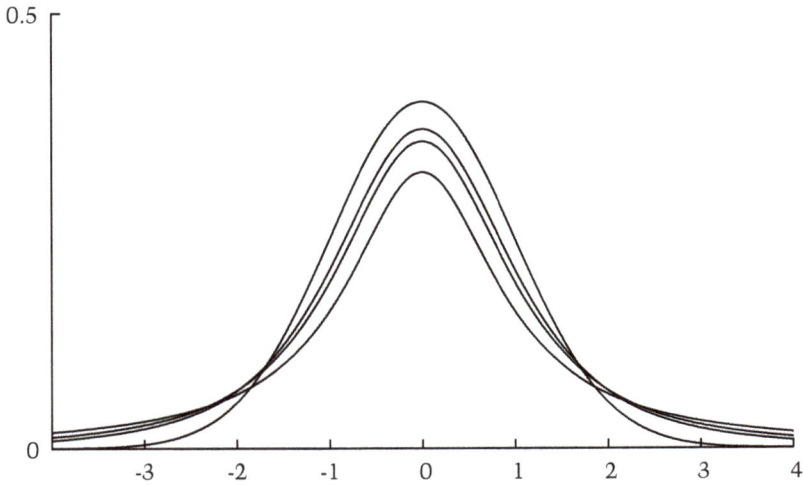

Figure 18: Student's t distributions (9.2): Cauchy $(k = 1)$, t_2 $(k = 2)$, t_3 $(k = 3)$, normal $(k \to \infty)$ (low to high peak).

'degrees of freedom'.

Student's t_2 (t_2) distribution [41] :

$$\mathrm{StudentsT}_2(x) = \frac{1}{(2 + x^2)^{\frac{3}{2}}} \tag{9.3}$$
$$= \mathrm{StudentsT}(x \; ; 2)$$
$$= \mathrm{PearsonVII}(x \; ; 0, \sqrt{2}, \tfrac{3}{2})$$

Student's t distribution with 2 degrees of freedom has a particularly simple form.

$$\mathrm{StudentsT}_2\mathrm{CDF}(x) = \tfrac{1}{2}\left(1 + \frac{x}{\sqrt{2 + x^2}}\right)$$

Table 9.1: Special cases of the Pearson type VII distribution

(9.1)	Pearson type VII	a	s	m
(9.2)	Student's t	0	$\sqrt{k}$	$\frac{k+1}{2}$
(9.3)	Student's t_2	0	$\sqrt{2}$	$\frac{3}{2}$
(9.4)	Student's t_3	0	$\sqrt{3}$	2
(9.5)	Student's z	0	1	$n/2$
(9.6)	Cauchy	.	.	1
(9.7)	standard Cauchy	0	1	1
(9.8)	relativistic Breit-Wigner	.	.	2

Student's t_3 (t_3) distribution [42] :

$$\mathrm{Students}T_3(x) = \frac{2}{\pi\left(1+\frac{x^2}{3}\right)^2} \tag{9.4}$$

$$= \mathrm{Students}T(x\ ;\ 3)$$
$$= \mathrm{RelBreitWigner}(x\ ;\ 0, \sqrt{3})$$
$$= \mathrm{PearsonVII}(x\ ;\ 0, \sqrt{3}, 2)$$

Student's t distribution with 3 degrees of freedom. Notable since the cumulative distribution function has a relatively simple form [42, p37].

$$\mathrm{Students}T_3\mathrm{CDF}(x) = \tfrac{1}{2} + \tfrac{1}{\sqrt{3}\pi}\left(\arctan(\tfrac{x}{\sqrt{3}}) + \tfrac{\frac{x}{\sqrt{3}}}{1+\frac{x^2}{3}}\right)$$

Student's z distribution [37, 39]:

$$\mathrm{Students}Z(z\ ;\ n) = \frac{1}{B(\frac{n-1}{2}, \frac{1}{2})}(1+z^2)^{-\frac{n}{2}} \tag{9.5}$$

$$= \mathrm{PearsonVII}(z\ ;\ 0, 1, \tfrac{n}{2})$$

The distribution of the statistic z, which was the original distribution investigated by Gosset (aka Student)[5] in his famous 1908 paper on the statis-

---

[5] Gosset's employer, the Guinness Brewing Company, insisted that he publish under a pseudonym.

tical error of sample means [37].

$$z = \frac{\bar{x} - \mu}{s}$$

$$\bar{x} = \frac{1}{n} \sum_{i=1}^{n} \text{Normal}_i(\mu, \sigma) ,$$

$$s^2 = \frac{1}{n} \sum_{i=1}^{n} \left(\text{Normal}_i(\mu, \sigma) - \bar{x}\right)^2$$

Here, $\bar{x}$ is the sample mean of n independent normal (4.1) random variables with mean μ and variance σ^2, and s^2 is the sample variance, except normalized by n rather than the now conventional $n - 1$. Latter work by Student and Fisher [38] resulted in a switch to the statistic $t = z/\sqrt{n-1}$.

Cauchy (Lorentz, Lorentzian, Cauchy-Lorentz, Breit-Wigner, normal ratio, Witch of Agnesi) distribution [43, 44, 3]:

$$\text{Cauchy}(x \; ; a, s) = \frac{1}{s\pi} \left(1 + \left(\frac{x-a}{s}\right)^2\right)^{-1} \tag{9.6}$$

$$= \text{PearsonVII}(x \; ; a, s, 1)$$

The Cauchy distribution is stable (21.20): a sum of independent Cauchy random variables is also Cauchy distributed.

$$\text{Cauchy}_1(a_1, s_1) + \text{Cauchy}_2(a_2, s_2) \sim \text{Cauchy}_3(a_1 + a_2, s_1 + s_2)$$

Standard Cauchy distribution [3]:

$$\text{StdCauchy}(x) = \frac{1}{\pi} \frac{1}{1 + x^2} \tag{9.7}$$

$$= \frac{1}{\pi}(x + i)^{-1}(x - i)^{-1}$$

$$= \text{Cauchy}(x \; ; 0, 1)$$

$$= \text{PearsonVII}(x \; ; 0, 1, 1)$$

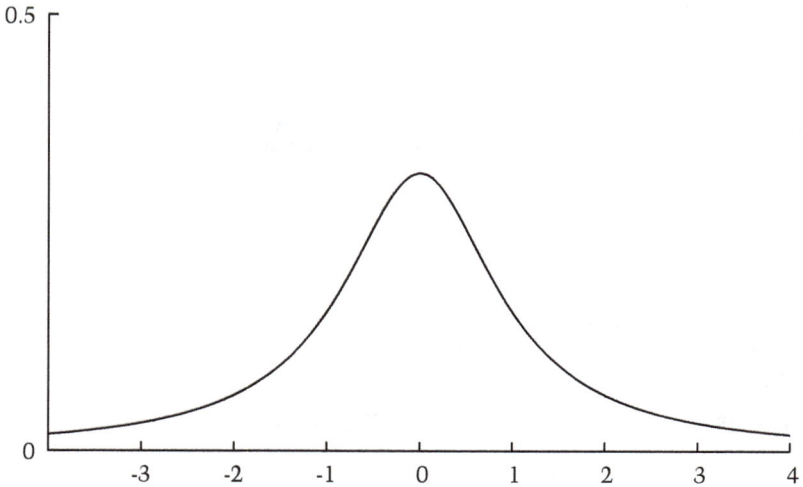

Figure 19: Standard Cauchy distribution, StdCauchy(x).

Relativistic Breit-Wigner (modified Lorentzian) distribution [45]:

$$\text{RelBreitWigner}(x \; ; \; a, s) = \frac{2}{|s|\pi}\left(1 + \left(\frac{x-a}{s}\right)^2\right)^{-2}$$

$$= \text{PearsonVII}(x \; ; \; a, s, 2)$$

Used to model the energy distribution of unstable particles in high-energy physics.

Interrelations

The Pearson VII distribution is a special case of the Pearson IV distribution (16.1). At high shape parameter m the Pearson VII limits to the normal distribution.

$$\text{Normal}(x \; ; \; \mu, \sigma) = \lim_{m\to\infty} \text{PearsonVII}(x \; ; \; \mu, \sigma\sqrt{2m}, m)$$

The Pearson type VII distribution is given by a ratio of normal and gamma random variables [42, p445].

$$\text{PearsonVII}(a, s, m) \sim a + s\sqrt{2m-1}\,\frac{\text{StdNormal}()}{\sqrt{\text{StdGamma}(m - \tfrac{1}{2})}}$$

The Cauchy distribution can be generated as a ratio of normal distributions

$$\text{Cauchy}(0, 1) \sim \frac{\text{Normal}_1(0, 1)}{\text{Normal}_2(0, 1)}$$

and as a ratio of gamma distributions [42, p427].

$$\left(\text{Cauchy}(0, 1)\right)^2 \sim \frac{\text{StdGamma}_1(\tfrac{1}{2})}{\text{StdGamma}_2(\tfrac{1}{2})}$$

Table 9.2: Properties of the Pearson VII distribution

Properties

notation	$\text{PearsonVII}(x \; ; a, s, m)$

PDF
$$\frac{1}{|s|B\left(m - \frac{1}{2}, \frac{1}{2}\right)}\left(1 + \left(\frac{x-a}{s}\right)^2\right)^{-m}$$

CDF / CCDF
$$\frac{1}{2} + \left(\frac{x-a}{s}\right)\frac{1}{B\left(m - \frac{1}{2}, \frac{1}{2}\right)} \, {}_2F_1\left(\frac{1}{2}, m; \frac{3}{2}; -\left(\frac{x-a}{s}\right)^2\right)$$

parameters	$a, \, s, \, m \in \mathbb{R}$	
	$m > \frac{1}{2}$	
support	$-\infty < x < +\infty$	
median	a	
mode	a	
mean	a	$m > 1$
variance	$\dfrac{s^2}{2m - 3}$	$m > \frac{3}{2}$
skew	0	$m > 2$
MGF	undefined	
CF	$e^{iat}\dfrac{2K_{m-\frac{1}{2}}(s|t|) \cdot \left(\frac{1}{2}s|t|\right)^{m-\frac{1}{2}}}{\Gamma\left(m - \frac{1}{2}\right)}$	$m > \frac{1}{2}$

10   Unit Gamma Distribution

Unit gamma (log-gamma, Grassia, log-Pearson III) distribution [46, 21, 47, 48]:

$$\mathrm{UnitGamma}(x \; ; \, a, s, \alpha, \beta) \tag{10.1}$$

$$= \frac{1}{\Gamma(\alpha)} \left| \frac{\beta}{s} \right| \left(\frac{x-a}{s} \right)^{\beta-1} \left(-\beta \ln \frac{x-a}{s} \right)^{\alpha-1}$$

$$\text{for } x, \; a, \; s, \; \alpha, \; \beta \text{ in } \mathbb{R}, \; \alpha > 0$$

$$\text{support } x \in [a, a+s], s > 0, \; \beta > 0$$

$$\text{or } x \in [a+s, a], s < 0, \; \beta > 0$$

$$\text{or } x \in [a+s, +\infty], s > 0, \; \beta < 0$$

$$\text{or } x \in [-\infty, a+s], s < 0, \; \beta < 0$$

A curious distribution that occurs as a limit of the generalized beta (17.1), and as the anti-log transform of the gamma distribution (7.1). For this reason, it is also sometimes called the log-gamma distribution.

Special cases

Uniform product distribution [49]:

$$\mathrm{UniformProduct}(x \; ; \, n) = \frac{1}{\Gamma(n)} (-\ln x)^{n-1} \tag{10.2}$$

$$= \mathrm{UnitGamma}(x \; ; \, 0, 1, n, 1)$$

$$0 > x > 1, \quad n = 1, \; 2, \; 3, \; \ldots$$

The product of n standard uniform distributions (1.2).

Interrelations

With $\alpha = 1$ we obtain the power function distribution (5.1) as a special case.

$$\mathrm{UnitGamma}(x \; ; \, a, s, 1, \beta) = \mathrm{PowerFn}(x \; ; \, a, s, \beta)$$

The unit gamma is the anti-log transform of the standard gamma distribution (7.2).

$$\text{UnitGamma}(0, 1, \alpha, \beta) \sim \exp(-\text{Gamma}(0, \tfrac{1}{\beta}, \alpha))$$

$$\text{UnitGamma}(0, 1, \alpha, 1) \sim \exp(-\text{StdGamma}(\alpha))$$

The unit gamma distribution is a limit of the generalized beta distribution (17.1), and limits to the gamma (7.1) and log-normal (6.1) [1] distributions.

$$\text{Gamma}(x \; ; \; a, s, \alpha) = \lim_{\beta \to \infty} \text{UnitGamma}(x \; ; \; a + \beta s, -\beta s, \alpha, \beta)$$

$$\lim_{\alpha \to \infty} \text{UnitGamma}(x \; ; \; a, \vartheta e^{\sigma \sqrt{\alpha}}, \alpha, \tfrac{\sqrt{\alpha}}{\sigma})$$

$$\propto \lim_{\alpha \to \infty} \left(\frac{x - a}{\vartheta e^{\sigma \sqrt{\alpha}}} \right)^{\frac{\sqrt{\alpha}}{\sigma} - 1} \left(-\frac{\sqrt{\alpha}}{\sigma} \ln \frac{x - a}{\vartheta e^{\sigma \sqrt{\alpha}}} \right)^{\alpha - 1}$$

$$\propto \left(\frac{x - a}{\vartheta} \right)^{-1} \lim_{\alpha \to \infty} \exp\left\{ \sqrt{\alpha} \frac{1}{\sigma} \ln \frac{x - a}{\vartheta} \right\} \left(1 - \frac{1}{\sqrt{\alpha}} \frac{1}{\sigma} \ln \frac{x - a}{\vartheta} \right)^{\alpha - 1}$$

$$\propto \left(\frac{x - a}{\vartheta} \right)^{-1} \lim_{\alpha \to \infty} e^{-z \sqrt{\alpha}} \left(1 + \frac{z}{\sqrt{\alpha}} \right)^{\alpha}, \quad z = -\tfrac{1}{\sigma} \ln \tfrac{x - a}{\vartheta}$$

$$\propto \left(\frac{x - a}{\vartheta} \right)^{-1} \exp\left\{ -\frac{1}{2\sigma^2} \left(\ln \frac{x - a}{\vartheta} \right)^2 \right\}$$

$$= \text{LogNormal}(x \; ; \; a, \vartheta, \sigma)$$

Here we utilize the Gaussian function limit $\lim_{c \to \infty} e^{-z \sqrt{c}} (1 + \frac{z}{\sqrt{c}})^c = e^{-\frac{1}{2} z^2}$ (§D).

The product of two unit-gamma distributions with common β is again a unit-gamma distribution [21, 1].

$$\text{UnitGamma}_1(0, s_1, \alpha_1, \beta) \; \text{UnitGamma}_2(0, s_2, \alpha_2, \beta)$$

$$\sim \text{UnitGamma}_3(0, s_1 s_2, \alpha_1 + \alpha_2, \beta)$$

The property is related to the analogous additive relation of the gamma

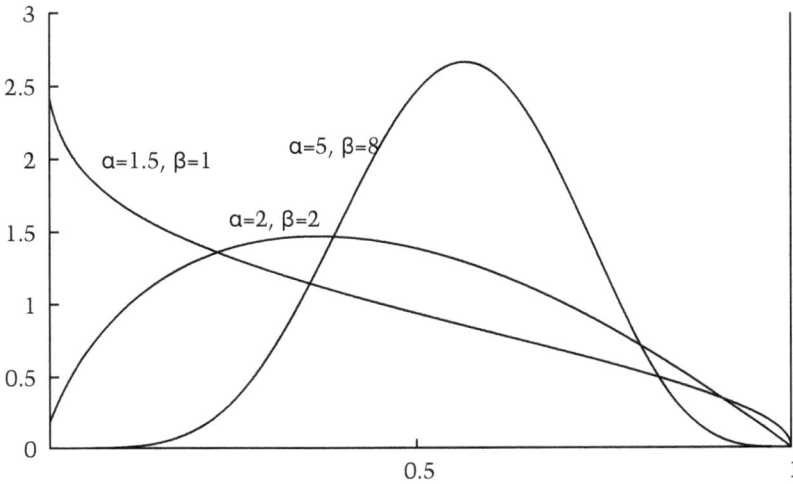

Figure 20: Unit gamma, finite support, $\mathrm{UnitGamma}(x\ ;0,1,\alpha,\beta)$, $\beta > 0$.

distribution.

$$\mathrm{UnitGamma}_1(0,s_1,\alpha_1,\beta)\ \mathrm{UnitGamma}_2(0,s_2,\alpha_2,\beta)$$

$$\sim s_1s_2(\mathrm{UnitGamma}_1(0,1,\alpha_1,1)\ \mathrm{UnitGamma}_2(0,1,\alpha_2,1))^{\frac{1}{\beta}}$$

$$\sim s_1s_2\left(e^{-\,\mathrm{StdGamma}_1(\alpha_1)-\mathrm{StdGamma}_2(\alpha_2)}\right)^{\frac{1}{\beta}}$$

$$\sim s_1s_2\left(e^{-\,\mathrm{StdGamma}_3(\alpha_1+\alpha_2)}\right)^{\frac{1}{\beta}}$$

$$\sim \mathrm{UnitGamma}_3(0,s_1s_2,\alpha_1+\alpha_2,\beta)$$

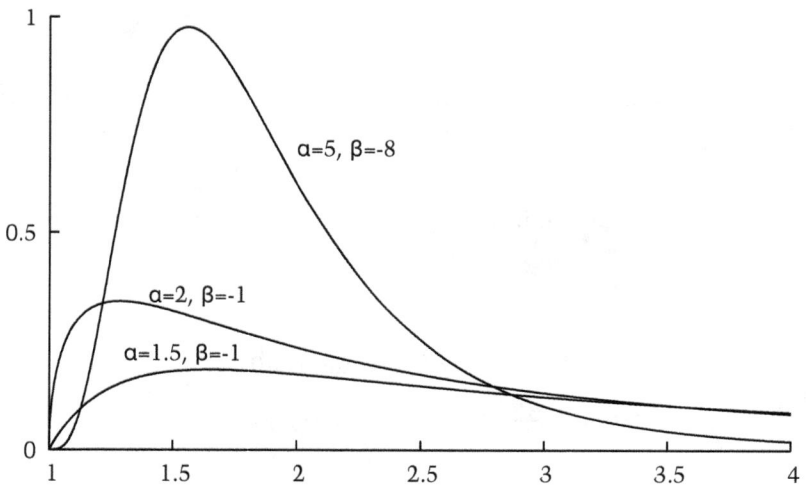

Figure 21: Unit gamma, semi-infinite support. $\text{UnitGamma}(x ; 0, 1, \alpha, \beta)$, $\beta < 0$

Table 10.1: Properties of the unit gamma distribution

Properties		
notation	$\mathrm{UnitGamma}(x \; ; a, s, \alpha, \beta)$	
PDF	$\dfrac{1}{\Gamma(\alpha)} \left| \dfrac{\beta}{s} \right| \left(\dfrac{x-a}{s} \right)^{\beta-1} \left(-\beta \ln \dfrac{x-a}{s} \right)^{\alpha-1}$	
CDF/CCDF	$1 - Q\left(\alpha, -\beta \ln \frac{x-a}{s}\right)$	$\frac{\beta}{s} > 0 \, / \, \frac{\beta}{s} < 0$
parameters	a, s, α, β in $\mathbb{R}$, $\alpha, \beta > 0$	
support	$[a, a+s], \; s > 0, \; \beta > 0$	
	$[a+s, a], \; s < 0, \; \beta > 0$	
	$[a+s, +\infty] \, s > 0, \; \beta < 0$	
	$[-\infty, a+s], \, s < 0, \; \beta < 0$	
mean	$a + s\left(\frac{\beta}{\beta+1} \right)^{\alpha}$	
variance	$s^2 \left(\frac{\beta}{\beta+2} \right)^{\alpha} - s^2 \left(\frac{\beta}{\beta+1} \right)^{2\alpha}$	
skew	not simple	
ex. kurtosis	not simple	
$E(X^h)$	$\left(\frac{\beta}{\beta+h} \right)^{\alpha}$	$a = 0 \, [47]$

11   Amoroso Distribution

The **Amoroso** (generalized gamma, Stacy-Mihram) distribution [50, 2, 31] is a four parameter, continuous, univariate, unimodal probability density, with semi-infinite support. The functional form in the most straightforward parameterization is

$$\text{Amoroso}(x \; ; a, \theta, \alpha, \beta) \tag{11.1}$$

$$= \frac{1}{\Gamma(\alpha)} \left| \frac{\beta}{\theta} \right| \left(\frac{x-a}{\theta} \right)^{\alpha\beta-1} \exp\left\{ -\left(\frac{x-a}{\theta} \right)^{\beta} \right\}$$

for x, a, θ, α, β in $\mathbb{R}$, $\alpha > 0$,

support $x \geqslant a$ if $\theta > 0$, $x \leqslant a$ if $\theta < 0$.

The Amoroso distribution was originally developed to model lifetimes [50]. It occurs as the Weibullization of the standard gamma distribution (7.1) and, with integer α, in extreme value statistics (11.24). The Amoroso distribution is itself a limiting form of various more general distributions, most notable the generalized beta (17.1) and generalized beta prime (18.1) distributions [51]. Many common and interesting probability distributions are special cases or limiting forms of the Amoroso (See Table 11.1).

The four real parameters of the Amoroso distribution consist of a location parameter a, a scale parameter θ, and two shape parameters, α and β. Whenever these symbols appears in special cases or limiting forms, they refer directly to the parameters of the Amoroso distribution. The shape parameter α is positive, and in many special cases an integer, $\alpha = n$, or half-integer, $\alpha = \frac{k}{2}$. The negation of a standard parameter is indicated by a bar, e.g. $\bar{\beta} = -\beta$. The chi, chi-squared and related distributions are traditionally parameterized with the scale parameter σ, where $\theta = (2\sigma^2)^{1/\beta}$, and σ is the standard deviation of a related normal distribution. Additional alternative parameters are introduced as necessary.

Special cases: Miscellaneous

The gamma distribution ($\beta = 1$) and it's special cases are detailed in (§7).

Stacy (anchored Amoroso, hyper gamma, generalized Weibull, Nukiyama-Tanasawa, generalized gamma, generalized semi-normal, Leonard hydrograph, hydrograph, transformed gamma) distribution [52, 53]:

$$\text{Stacy}(x \; ; \theta, \alpha, \beta) = \frac{1}{\Gamma(\alpha)} \left| \frac{\beta}{\theta} \right| \left(\frac{x}{\theta} \right)^{\alpha\beta-1} \exp\left\{ -\left(\frac{x}{\theta} \right)^{\beta} \right\} \qquad (11.2)$$

$$= \text{Amoroso}(x \; ; 0, \theta, \alpha, \beta)$$

If we drop the location parameter from Amoroso, then we obtain the Stacy, or generalized gamma distribution. If β is negative then the distribution is **generalized inverse gamma**, the parent of various inverse distributions, including the inverse gamma (11.13) and inverse chi (11.19).

The Stacy distribution is obtained as the positive even powers, modulus, and powers of the modulus of a centered, normal random variable (4.1),

$$\text{Stacy}\left((2\sigma^2)^{\frac{1}{\beta}}, \tfrac{1}{2}, \beta \right) \sim \left| \text{Normal}(0, \sigma) \right|^{\frac{2}{\beta}}$$

and as powers of the sum of squares of k centered, normal random variables.

$$\text{Stacy}\left((2\sigma^2)^{\frac{1}{\beta}}, \tfrac{1}{2}k, \beta \right) \sim \left(\sum_{i=1}^{k} \left(\text{Normal}(0, \sigma) \right)^2 \right)^{\frac{1}{\beta}}$$

Pseudo-Weibull distribution [54]:

$$\text{PseudoWeibull}(x \; ; a, \theta, \beta) = \frac{1}{\Gamma(1 + \frac{1}{\beta})} \frac{\beta}{|\theta|} \left(\frac{x-a}{\theta} \right)^{\beta} \exp\left\{ -\left(\frac{x-a}{\theta} \right)^{\beta} \right\}$$

$$(11.3)$$

$$\text{for } \beta > 0$$

$$= \text{Amoroso}(x \; ; a, \theta, 1 + \tfrac{1}{\beta}, \beta)$$

Proposed as another model of failure times.

Table 11.1: Special cases of the Amoroso family

(11.1)	Amoroso	a	θ	α	β
(11.2)	Stacy	0	.	.	.
(11.4)	half exponential power	.	.	$\frac{1}{\beta}$	.
(11.24)	gen. Fisher-Tippett	.	.	n	.
(11.25)	Fisher-Tippett	.	.	1	.
(11.29)	Fréchet	.	.	1	<0
(11.28)	generalized Fréchet	.	.	n	<0
(11.23)	inverse Nakagami	.	.	.	-2
(11.18)	scaled inverse chi	0	.	$\frac{1}{2}k$	-2
(11.19)	inverse chi	0	$\frac{1}{\sqrt{2}}$	$\frac{1}{2}k$	-2
(11.21)	inverse Maxwell	0	.	$\frac{3}{2}$	-2
(11.20)	inverse Rayleigh	0	.	1	-2
(11.22)	inverse half normal	0	.	$\frac{1}{2}$	-2
(11.13)	inverse gamma	.	.	.	-1
(11.16)	scaled inverse chi-square	0	.	$\frac{1}{2}k$	-1
(11.17)	inverse chi-square	0	$\frac{1}{2}$	$\frac{1}{2}k$	-1
(11.15)	Lévy	.	.	$\frac{1}{2}$	-1
(11.14)	inverse exponential	0	.	1	-1
(7.1)	gamma	.	.	.	1
(11.5)	Hohlfeld	0	.	$\frac{2}{3}$	$\frac{3}{2}$
(11.6)	Nakagami	.	.	.	2
(11.9)	scaled chi	0	.	$\frac{1}{2}k$	2
(11.8)	chi	0	$\sqrt{2}$	$\frac{1}{2}k$	2
(11.7)	half normal	0	.	$\frac{1}{2}$	2
(11.10)	Rayleigh	0	.	1	2
(11.11)	Maxwell	0	.	$\frac{3}{2}$	2
(11.12)	Wilson-Hilferty	0	.	.	3
(11.26)	generalized Weibull	.	.	n	>0
(11.27)	Weibull	.	.	1	>0
(11.3)	pseudo-Weibull	.	.	$1+\frac{1}{\beta}$	>0

(k, n positive integers)

For special cases of the gamma distribution ($\beta = 1$) see table 7.1.

Half exponential power (half Subbotin) distribution [55]:

$$\mathrm{HalfExpPower}(x \; ; \; a, \theta, \beta) = \frac{1}{\Gamma(\frac{1}{\beta})} \left| \frac{\beta}{\theta} \right| \exp\left\{ -\left(\frac{x-a}{\theta} \right)^\beta \right\} \qquad (11.4)$$

$$= \mathrm{Amoroso}(x \; ; \; a, \theta, \tfrac{1}{\beta}, \beta)$$

As the name implies, half an exponential power (21.4) distribution. Special cases include $\beta = -1$ inverse exponential (11.14), $\beta = 1$ exponential (2.1), $\beta = \frac{2}{3}$ Hohlfeld (11.5) and $\beta = 2$ half normal (11.7) distributions.

Hohlfeld distribution [56]:

$$\mathrm{Hohlfeld}(x \; ; \; a, \theta) = \frac{1}{\Gamma(\frac{2}{3})} \left| \frac{3}{2\theta} \right| \exp\left\{ -\left(\frac{x-a}{\theta} \right)^{3/2} \right\} \qquad (11.5)$$

$$= \mathrm{HalfExpPower}(x \; ; \; a, \theta, \tfrac{3}{2})$$

$$= \mathrm{Amoroso}(x \; ; \; a, \theta, \tfrac{2}{3}, \tfrac{3}{2})$$

Occurs in the extreme statistics of Brownian ratchets [56, Suppl. p.5].

Special cases: Positive integer β

With $\beta = 1$ we obtain the gamma family of distributions: gamma (7.1), standard gamma (7.2) and chi square (7.3) distributions. See (§7).

Nakagami (generalized normal, Nakagami-m, m) distribution [57]:

$$\mathrm{Nakagami}(x \; ; \; a, \theta, \alpha) \qquad (11.6)$$

$$= \frac{2}{\Gamma(\alpha)|\theta|} \left(\frac{x-a}{\theta} \right)^{2\alpha-1} \exp\left\{ -\left(\frac{x-a}{\theta} \right)^2 \right\}$$

$$= \mathrm{Amoroso}(x \; ; \; a, \theta, \alpha, 2)$$

Used to model attenuation of radio signals that reach a receiver by multiple paths [57].

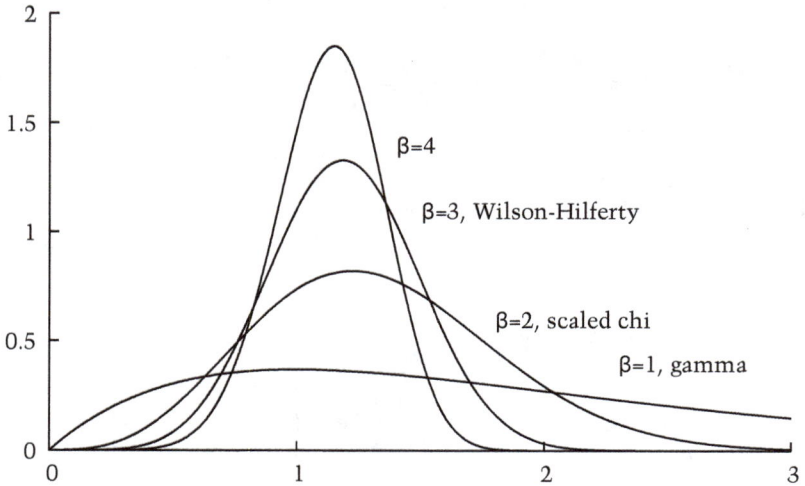

Figure 22: Gamma, scaled chi, and Wilson-Hilferty distributions, Amoroso$(x\ ;\ 0, 1, 2, \beta)$

Half normal (semi-normal, positive definite normal, one-sided normal) distribution [2]:

$$\mathrm{HalfNormal}(x\ ;\ a, \sigma) = \frac{2}{\sqrt{2\pi\sigma^2}} \exp\left\{ -\left(\frac{(x-a)^2}{2\sigma^2} \right) \right\} \qquad (11.7)$$

$$(x-a)/\sigma > 0$$

$$= \mathrm{Amoroso}(x\ ;\ a, \sqrt{2\sigma^2}, \tfrac{1}{2}, 2)$$

The modulus of a normal distribution about the mean.

Chi (χ) distribution [2]:

$$\text{Chi}(x \; ; \; k) = \frac{\sqrt{2}}{\Gamma(\frac{k}{2})} \left(\frac{x}{\sqrt{2}} \right)^{k-1} \exp\left\{ -\left(\frac{x^2}{2} \right) \right\} \tag{11.8}$$

for positive integer k

$$= \text{ScaledChi}(x \; ; \; 1, k)$$

$$= \text{Stacy}(x \; ; \; \sqrt{2}, \tfrac{k}{2}, 2)$$

$$= \text{Amoroso}(x \; ; \; 0, \sqrt{2}, \tfrac{k}{2}, 2)$$

The root-mean-square of k independent standard normal variables, or the square root of a chi-square random variable.

$$\text{Chi}(k) \sim \sqrt{\text{ChiSqr}(k)}$$

Scaled chi (generalized Rayleigh) distribution [58, 2]:

$$\text{ScaledChi}(x \; ; \; \sigma, k) = \frac{2}{\Gamma(\frac{k}{2})\sqrt{2\sigma^2}} \left(\frac{x}{\sqrt{2\sigma^2}} \right)^{k-1} \exp\left\{ -\left(\frac{x^2}{2\sigma^2} \right) \right\}$$

for positive integer k

$$= \text{Stacy}(x \; ; \; \sqrt{2\sigma^2}, \tfrac{k}{2}, 2) \tag{11.9}$$

$$= \text{Amoroso}(x \; ; \; 0, \sqrt{2\sigma^2}, \tfrac{k}{2}, 2)$$

The root-mean-square of k independent and identically distributed normal variables with zero mean and variance σ^2.

Rayleigh (circular normal) distribution [59, 2]:

$$\text{Rayleigh}(x \; ; \; \sigma) = \frac{1}{\sigma^2} \, x \, \exp\left\{ -\left(\frac{x^2}{2\sigma^2} \right) \right\} \tag{11.10}$$

$$= \text{ScaledChi}(x \; ; \; \sigma, 2)$$

$$= \text{Stacy}(x \; ; \; \sqrt{2\sigma^2}, 1, 2)$$

$$= \text{Amoroso}(x \; ; \; 0, \sqrt{2\sigma^2}, 1, 2)$$

The root-mean-square of two independent and identically distributed normal variables with zero mean and variance σ^2. For instance, wind speeds are approximately Rayleigh distributed, since the horizontal components

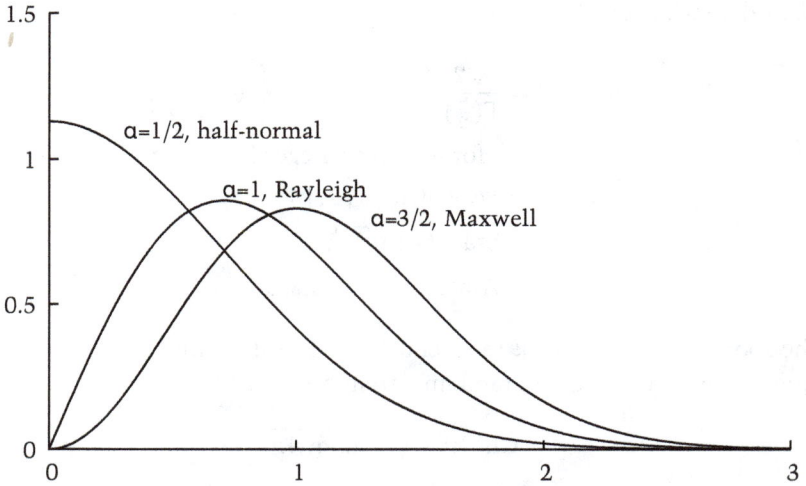

Figure 23: Half normal, Rayleigh, and Maxwell distributions, Amoroso(x ; 0, 1, α, 2)

of the velocity are approximately normal, and the vertical component is typically small [60].

Maxwell (Maxwell-Boltzmann, Maxwell speed, spherical normal) distribution [61, 62]:

$$\text{Maxwell}(x \ ; \ \sigma) = \frac{\sqrt{2}}{\sqrt{\pi}\sigma^3} x^2 \exp\left\{-\left(\frac{x^2}{2\sigma^2}\right)\right\} \qquad (11.11)$$

$$= \text{ScaledChi}(x \ ; \ \sigma, 3)$$

$$= \text{Stacy}(x \ ; \ \sqrt{2\sigma^2}, \tfrac{3}{2}, 2)$$

$$= \text{Amoroso}(x \ ; \ 0, \sqrt{2\sigma^2}, \tfrac{3}{2}, 2)$$

The speed distribution of molecules in thermal equilibrium. The root-mean-square of three independent and identically distributed normal variables with zero mean and variance σ^2.

Wilson-Hilferty distribution [63, 2]:

$$\mathrm{WilsonHilferty}(x \; ; \theta, \alpha) = \frac{3}{\Gamma(\alpha)|\theta|}\left(\frac{x}{\theta}\right)^{3\alpha-1} \exp\left\{-\left(\frac{x}{\theta}\right)^3\right\} \qquad (11.12)$$

$$= \mathrm{Stacy}(x \; ; \theta, \alpha, 3)$$

$$= \mathrm{Amoroso}(x \; ; 0, \theta, \alpha, 3)$$

The cube root of a gamma variable follows the Wilson-Hilferty distribution [63], which has been used to approximate a normal distribution if α is not too small.

$$\mathrm{WilsonHilferty}(x \; ; \theta, \alpha) \approx \mathrm{Normal}(x \; ; 1 - \tfrac{2}{9\alpha}, \tfrac{2}{9\alpha})$$

A related approximation using quartic roots of gamma variables [64] leads to $\mathrm{Amoroso}(x \; ; 0, \theta, \alpha, 4)$.

Special cases: Negative integer β

With negative β we obtain various "inverse" distributions related to distributions with positive β by the reciprocal transformation $(\frac{x-a}{\theta}) \to (\frac{\theta}{x-a})$.

Inverse gamma (Pearson type V, March, Vinci) distribution [6, 2]:

$$\mathrm{InvGamma}(x \; ; \theta, \alpha) = \frac{1}{\Gamma(\alpha)|\theta|}\left(\frac{\theta}{x - a}\right)^{\alpha+1} \exp\left\{-\left(\frac{\theta}{x - a}\right)\right\} \qquad (11.13)$$

$$= \mathrm{Amoroso}(x \; ; a, \theta, \alpha, -1)$$

Occurs as the conjugate prior for an exponential distribution's scale parameter [2], or the prior for variance of a normal distribution with known mean [65]. Frequently defined with zero scale parameter.

Inverse exponential distribution [66]:

$$\mathrm{InvExp}(x \; ; a, \theta) = \frac{1}{|\theta|}\left(\frac{\theta}{x - a}\right)^2 \exp\left\{-\left(\frac{\theta}{x - a}\right)\right\} \qquad (11.14)$$

$$= \mathrm{InvGamma}(x \; ; a, \theta, 1)$$

$$= \mathrm{Amoroso}(x \; ; a, \theta, 1, -1)$$

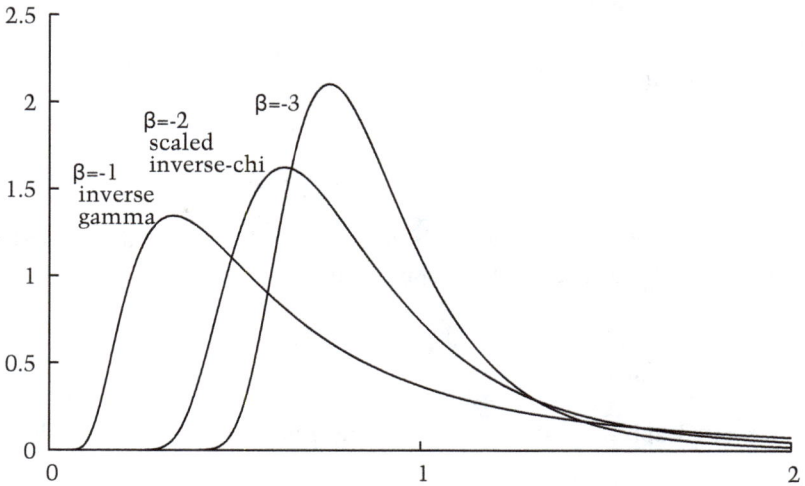

Figure 24: Inverse gamma and scaled inverse-chi distributions, Amoroso(x ; $0, 1, 2, \beta$), negative β.

Note that the name "inverse exponential" is occasionally used for the ordinary exponential distribution (2.1).

Lévy distribution (van der Waals profile) [67]:

$$\text{Lévy}(x \; ; \; a, c) = \sqrt{\frac{|c|}{2\pi}} \frac{1}{(x-a)^{3/2}} \exp\left\{-\frac{c}{2(x-a)}\right\} \qquad (11.15)$$
$$= \text{Amoroso}(x \; ; \; a, \tfrac{c}{2}, \tfrac{1}{2}, -1)$$

The Lévy distribution is notable for being stable: a linear combination of identically distributed Lévy distributions is again a Lévy distribution. The other stable distributions with analytic forms are the normal distribution (4.1), which is also a limit of the Amoroso distribution, and the Cauchy distribution (9.6), which is not. Lévy distributions describe first passage times in one dimension [67]. See also the inverse Gaussian distribution (20.3), the first passage time distribution for Brownian diffusion with drift.

Scaled inverse chi-square distribution [65]:

$$\text{ScaledInvChiSqr}(x\,;\,\sigma,k) \tag{11.16}$$

$$= \frac{2\sigma^2}{\Gamma(\frac{k}{2})}\left(\frac{1}{2\sigma^2 x}\right)^{\frac{k}{2}+1}\exp\left\{-\left(\frac{1}{2\sigma^2 x}\right)\right\}$$

$$\text{for positive integer } k$$

$$= \text{InvGamma}(x\,;\,0,\tfrac{1}{2\sigma^2},\tfrac{k}{2})$$

$$= \text{Stacy}(x\,;\,\tfrac{1}{2\sigma^2},\tfrac{k}{2},-1)$$

$$= \text{Amoroso}(x\,;\,0,\tfrac{1}{2\sigma^2},\tfrac{k}{2},-1)$$

A special case of the inverse gamma distribution with half-integer α. Used as a prior for variance parameters in normal models [65].

Inverse chi-square distribution [65]:

$$\text{InvChiSqr}(x\,;\,k) = \frac{2}{\Gamma(\frac{k}{2})}\left(\frac{1}{2x}\right)^{\frac{k}{2}+1}\exp\left\{-\left(\frac{1}{2x}\right)\right\} \tag{11.17}$$

$$\text{for positive integer } k$$

$$= \text{ScaledInvChiSqr}(x\,;\,1,k)$$

$$= \text{InvGamma}(x\,;\,0,\tfrac{1}{2},\tfrac{k}{2})$$

$$= \text{Stacy}(x\,;\,\tfrac{1}{2},\tfrac{k}{2},-1)$$

$$= \text{Amoroso}(x\,;\,0,\tfrac{1}{2},\tfrac{k}{2},-1)$$

A standard scaled inverse chi-square distribution.

Scaled inverse chi distribution [27]:

$$\text{ScaledInvChi}(x\,;\,\sigma,k) \tag{11.18}$$

$$= \frac{2\sqrt{2\sigma^2}}{\Gamma(\frac{k}{2})}\left(\frac{1}{\sqrt{2\sigma^2}x}\right)^{k+1}\exp\left\{-\left(\frac{1}{2\sigma^2 x^2}\right)\right\}$$

$$= \text{Stacy}(x\,;\,\tfrac{1}{\sqrt{2\sigma^2}},\tfrac{k}{2},-2)$$

$$= \text{Amoroso}(x\,;\,0,\tfrac{1}{\sqrt{2\sigma^2}},\tfrac{k}{2},-2)$$

Used as a prior for the standard deviation of a normal distribution.

Inverse chi distribution [27]:

$$\text{InvChi}(x \; ; \; k) = \frac{2\sqrt{2}}{\Gamma(\frac{k}{2})} \left(\frac{1}{\sqrt{2}x} \right)^{k+1} \exp\left\{ -\left(\frac{1}{2x^2} \right) \right\} \quad (11.19)$$
$$= \text{Stacy}(x \; ; \; \tfrac{1}{\sqrt{2}}, \tfrac{k}{2}, -2)$$
$$= \text{Amoroso}(x \; ; \; 0, \tfrac{1}{\sqrt{2}}, \tfrac{k}{2}, -2)$$

Inverse Rayleigh distribution [68]:

$$\text{InvRayleigh}(x \; ; \; \sigma) = 2\sqrt{2\sigma^2} \left(\frac{1}{\sqrt{2\sigma^2}x} \right)^{3} \exp\left\{ -\left(\frac{1}{2\sigma^2 x^2} \right) \right\} \quad (11.20)$$
$$= \text{Stacy}(x \; ; \; \tfrac{1}{\sqrt{2\sigma^2}}, 1, -2)$$
$$= \text{Fréchet}(x \; ; \; 0, \tfrac{1}{\sqrt{2\sigma^2}}, 2)$$
$$= \text{Amoroso}(x \; ; \; 0, \tfrac{1}{\sqrt{2\sigma^2}}, 1, -2)$$

The inverse Rayleigh distribution has been used to model failure time [69].

Inverse Maxwell distribution [70]:

$$\text{InvMaxwell}(x \; ; \; \sigma) = \frac{\sqrt{2\sigma^2}}{\sqrt{\pi}} \left(\frac{1}{\sqrt{2\sigma^2}x} \right)^{4} \exp\left\{ -\left(\frac{1}{2\sigma^2 x^2} \right) \right\} \quad (11.21)$$
$$= \text{ScaledInvChi}(x \; ; \; \sigma, 3)$$
$$= \text{Amoroso}(x \; ; \; 0, \tfrac{1}{\sqrt{2\sigma^2}}, \tfrac{3}{2}, -2)$$

Inverse half-normal distribution [70]:

$$\text{InvHalfNormal}(x \; ; \; a, \sigma) = \frac{2}{\sqrt{2\sigma^2}} \frac{1}{(x-a)^2} \exp\left\{ -\left(\frac{1}{2\sigma^2(x-a)^2} \right) \right\} \quad (11.22)$$
$$= \text{Amoroso}(x \; ; \; a, \tfrac{1}{\sqrt{2\sigma^2}}, \tfrac{1}{2}, -2)$$

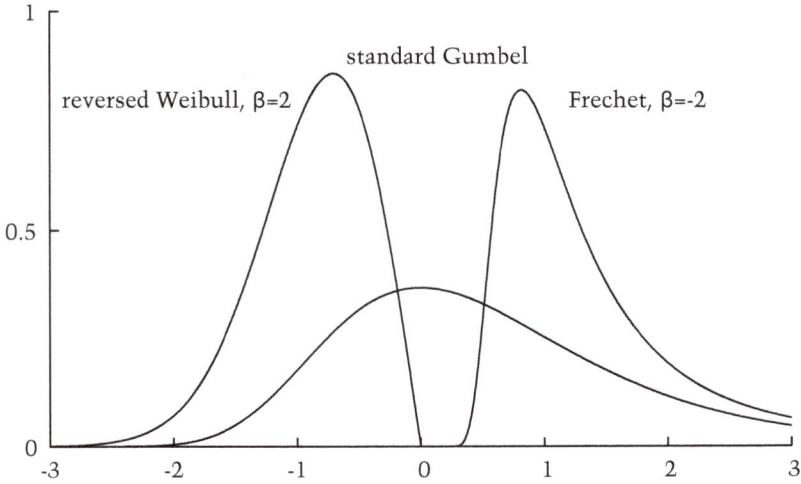

Figure 25: Extreme value distributions of maxima.

Inverse Nakagami distribution [71]:

$$\mathrm{InvNakagami}(x \; ; \; a, \theta, \alpha) \tag{11.23}$$

$$= \frac{2}{\Gamma(\alpha)|\theta|} \left(\frac{\theta}{x-a}\right)^{2\alpha+1} \exp\left\{-\left(\frac{\theta}{x-a}\right)^2\right\}$$

$$= \mathrm{Amoroso}(x \; ; \; a, \theta, \alpha, -2)$$

Special cases: Extreme order statistics

Generalized Fisher-Tippett distribution [72, 73]:

$$\mathrm{GenFisherTippett}(x \; ; \; a, \omega, n, \beta)$$

$$= \frac{n^n}{\Gamma(n)} \left|\frac{\beta}{\omega}\right| \left(\frac{x-a}{\omega}\right)^{n\beta-1} \exp\left\{-n\left(\frac{x-a}{\omega}\right)^\beta\right\}$$

$$\text{for positive integer } n \tag{11.24}$$

$$= \mathrm{Amoroso}(x \; ; \; a, \omega/n^{\frac{1}{\beta}}, n, \beta)$$

If we take N samples from a probability distribution, then asymptotically for large N and $n \ll N$, the distribution of the nth largest (or smallest) sample follows a generalized Fisher-Tippett distribution. The parameter β depends on the tail behavior of the sampled distribution. Roughly speaking, if the tail is unbounded and decays exponentially then β limits to ∞, if the tail scales as a power law then $\beta < 0$, and if the tail is finite $\beta > 0$ [32]. In these three limits we obtain the Gumbel (8.5, 8.4), Fréchet (11.29, 11.28) and Weibull (11.27,11.26) families of extreme value distribution (Extreme value distributions types I, II and III) respectively. If β/ω is negative we obtain distributions for the nth maxima, if positive then the nth minima.

Fisher-Tippett (Generalized extreme value, GEV, von Mises-Jenkinson, von Mises extreme value, log-Gumbel, Brody) distribution [33, 74, 32, 3, 75]:

$$\text{FisherTippett}(x \; ; a, \omega, \beta) \tag{11.25}$$

$$= \left| \frac{\beta}{\omega} \right| \left(\frac{x-a}{\omega} \right)^{\beta-1} \exp\left\{ -\left(\frac{x-a}{\omega} \right)^{\beta} \right\}$$

$$= \text{GenFisherTippett}(x \; ; a, \omega, 1, \beta)$$

$$= \text{Amoroso}(x \; ; a, \omega, 1, \beta)$$

The asymptotic distribution of the extreme value from a large sample. The superclass of type I, II and III (Gumbel, Fréchet, Weibull) extreme value distributions [74]. This is the **max stable distribution** (distribution of maxima) with $\beta/\omega < 0$ and the **min stable distribution** (distribution of minima) for $\beta/\omega > 0$.

The maximum of two Fisher-Tippett random variables (minimum if $\beta/\omega > 0$) is again a Fisher-Tippett random variable.

$$\max \left[\text{FisherTippett}(a, \omega_1, \beta), \text{FisherTippett}(a, \omega_2, \beta) \right]$$

$$\sim \text{FisherTippett}(a, \frac{\omega_1 \omega_2}{(\omega_1^\beta + \omega_2^\beta)^{1/\beta}}, \beta)$$

This follows since taking the maximum of two random variables is equivalent to multiplying their cumulative distribution functions, and the Fisher-Tippett cumulative distribution function is $\exp\left\{ -\left(\frac{x-a}{\omega} \right)^{\beta} \right\}$.

Generalized Weibull distribution [72, 73]:

$$\mathrm{GenWeibull}(x\ ;\ a, \omega, n, \beta) \hspace{4cm} (11.26)$$

$$= \frac{n^n}{\Gamma(n)} \frac{\beta}{|\omega|} \left(\frac{x-a}{\omega}\right)^{n\beta-1} \exp\left\{-n\left(\frac{x-a}{\omega}\right)^{\beta}\right\}$$

$$\text{for } \beta > 0$$

$$= \mathrm{GenFisherTippett}(x\ ;\ a, \omega, n, \beta)$$

$$= \mathrm{Amoroso}(x\ ;\ a, \omega/n^{\frac{1}{\beta}}, n, \beta)$$

The limiting distribution of the nth smallest value of a large number of identically distributed random variables that are at least a. If ω is negative we obtain the distribution of the nth largest value.

Weibull (Fisher-Tippett type III, Gumbel type III, Rosin-Rammler, Rosin-Rammler-Weibull, extreme value type III, Weibull-Gnedenko, stretched exponential) distribution [76, 3]:

$$\mathrm{Weibull}(x\ ;\ a, \omega, \beta) = \frac{\beta}{|\omega|} \left(\frac{x-a}{\omega}\right)^{\beta-1} \exp\left\{-\left(\frac{x-a}{\omega}\right)^{\beta}\right\} \hspace{1cm} (11.27)$$

$$\text{for } \beta > 0$$

$$= \mathrm{FisherTippett}(x\ ;\ a, \omega, \beta)$$

$$= \mathrm{Amoroso}(x\ ;\ a, \omega, 1, \beta)$$

Weibull[6] is the limiting distribution of the minimum of a large number of identically distributed random variables that are at least a. If ω is negative we obtain a **reversed Weibull** (extreme value type III) distribution for maxima. Special cases of the Weibull distribution include the exponential ($\beta = 1$) and Rayleigh ($\beta = 2$) distributions.

---

[6]Pronounced variously as *vay-bull* or *wye-bull*.

Generalized Fréchet distribution [72, 73]:

$$\text{GenFréchet}(x \; ; \; a, \omega, n, \bar{\beta}) \tag{11.28}$$

$$= \frac{n^n}{\Gamma(n)} \frac{\bar{\beta}}{|\omega|} \left(\frac{x-a}{\omega} \right)^{-n\bar{\beta}-1} \exp\left\{ -n \left(\frac{x-a}{\omega} \right)^{-\bar{\beta}} \right\}$$

$$\text{for } \bar{\beta} > 0$$

$$= \text{GenFisherTippett}(x \; ; \; a, \omega, n, -\bar{\beta})$$

$$= \text{Amoroso}(x \; ; \; a, \omega/n^{\frac{1}{\beta}}, n, -\bar{\beta}),$$

The limiting distribution of the nth largest value of a large number of identically distributed random variables whose moments are not all finite (i.e. heavy tailed distributions). (If the shape parameter ω is negative then minimum rather than maxima.)

Fréchet (extreme value type II, Fisher-Tippett type II, Gumbel type II, inverse Weibull) distribution [77, 32]:

$$\text{Fréchet}(x \; ; \; a, \omega, \bar{\beta}) = \frac{\bar{\beta}}{|\omega|} \left(\frac{x-a}{\omega} \right)^{-\bar{\beta}-1} \exp\left\{ -\left(\frac{x-a}{\omega} \right)^{-\bar{\beta}} \right\} \tag{11.29}$$

$$\text{for } \bar{\beta} > 0$$

$$= \text{FisherTippett}(x \; ; \; a, \omega, -\bar{\beta})$$

$$= \text{Amoroso}(x \; ; \; a, \omega, 1, -\bar{\beta})$$

The limiting distribution of the maximum of a large number of identically distributed random variables whose moments are not all finite (i.e. heavy tailed distributions). (If the shape parameter ω is negative then minimum rather than maxima.) Special cases of the Fréchet distribution include the inverse exponential ($\bar{\beta} = 1$) and inverse Rayleigh ($\bar{\beta} = 2$) distributions.

Interrelations

The Amoroso distribution is a limiting form of the generalized beta (17.1) and generalized beta prime (18.1) distributions [51]. Limits of the Amoroso distribution include gamma-exponential (8.1), log-normal (6.1), and normal

Table 11.2: Properties of the Amoroso distribution

Properties

notation	$\text{Amoroso}(x \; ; \; a, \theta, \alpha, \beta)$	
PDF	$\dfrac{1}{\Gamma(\alpha)} \left| \dfrac{\beta}{\theta} \right| \left(\dfrac{x-a}{\theta} \right)^{\alpha\beta-1} \exp\left\{ -\left(\dfrac{x-a}{\theta} \right)^{\beta} \right\}$	
CDF / CCDF	$1 - Q\left(\alpha, \left(\frac{x-a}{\theta} \right)^{\beta} \right)$	$\frac{\theta}{\beta} > 0 \,/\, \frac{\theta}{\beta} < 0$
parameters	a, θ, α, β in $\mathbb{R}$, $\alpha > 0$	
support	$x \geqslant a$	$\theta > 0$
	$x \leqslant a$	$\theta < 0$
mode	$a + \theta(\alpha - \frac{1}{\beta})^{\frac{1}{\beta}}$	$\alpha\beta \geqslant 1$
	a	$\alpha\beta \leqslant 1$
mean	$a + \theta \dfrac{\Gamma(\alpha + \frac{1}{\beta})}{\Gamma(\alpha)}$	$\alpha + \frac{1}{\beta} \geqslant 0$
variance	$\theta^2 \left[\dfrac{\Gamma(\alpha + \frac{2}{\beta})}{\Gamma(\alpha)} - \dfrac{\Gamma(\alpha + \frac{1}{\beta})^2}{\Gamma(\alpha)^2} \right]$	$\alpha + \frac{2}{\beta} \geqslant 0$
skew	$\text{sgn}\left(\frac{\beta}{\theta} \right) \left[\dfrac{\Gamma(\alpha+\frac{3}{\beta})}{\Gamma(\alpha)} - 3 \dfrac{\Gamma(\alpha+\frac{2}{\beta})\Gamma(\alpha+\frac{1}{\beta})}{\Gamma(\alpha)^2} + 2 \dfrac{\Gamma(\alpha+\frac{1}{\beta})^3}{\Gamma(\alpha)^3} \right]$ $\bigg/ \left[\dfrac{\Gamma(\alpha+\frac{2}{\beta})}{\Gamma(\alpha)} - \dfrac{\Gamma(\alpha+\frac{1}{\beta})^2}{\Gamma(\alpha)^2} \right]^{3/2}$	
ex. kurtosis	$\left[\dfrac{\Gamma(\alpha+\frac{4}{\beta})}{\Gamma(\alpha)} - 4 \dfrac{\Gamma(\alpha+\frac{3}{\beta})\Gamma(\alpha+\frac{1}{\beta})}{\Gamma(\alpha)^2} + 6 \dfrac{\Gamma(\alpha+\frac{2}{\beta})\Gamma(\alpha+\frac{1}{\beta})^2}{\Gamma(\alpha)^3} \right.$ $\left. - 3 \dfrac{\Gamma(\alpha+\frac{1}{\beta})^4}{\Gamma(\alpha)^4} \right] \bigg/ \left[\dfrac{\Gamma(\alpha+\frac{2}{\beta})}{\Gamma(\alpha)} - \dfrac{\Gamma(\alpha+\frac{1}{\beta})^2}{\Gamma(\alpha)^2} \right]^2 - 3$	
entropy	$\ln \dfrac{|\theta|\Gamma(\alpha)}{|\beta|} + \alpha + \left(\frac{1}{\beta} - \alpha \right) \psi(\alpha)$	[53]

(4.1) [2] and power function (5.1) distributions.

$$\text{GammaExp}(x \; ; \; \nu, \lambda, \alpha) = \lim_{\beta \to \infty} \text{Amoroso}(x \; ; \; \nu + \beta\lambda, -\beta\lambda, \alpha, \beta)$$

$$\text{LogNormal}(x \; ; \; a, \vartheta, \sigma) = \lim_{\alpha \to \infty} \text{Amoroso}(x \; ; \; a, \vartheta\alpha^{-\sigma\sqrt{\alpha}}, \alpha, \tfrac{1}{\sigma\sqrt{\alpha}})$$

$$\text{Normal}(x \; ; \; \mu, \sigma) = \lim_{\alpha \to \infty} \text{Amoroso}(x \; ; \; 0, \mu - \sigma\sqrt{\alpha}, \tfrac{\sigma}{\sqrt{\alpha}}, \alpha, 1)$$

The log-normal limit is particularly subtle [78], (§D).

$$\lim_{\alpha \to \infty} \text{Amoroso}(x \; ; \; a, \vartheta\alpha^{-\sigma\sqrt{\alpha}}, \alpha, \tfrac{1}{\sigma\sqrt{\alpha}})$$

Ignore normalization constants and rearrange,

$$\propto \left(\tfrac{x-a}{\vartheta}\right)^{-1} \exp\left\{ \alpha \ln(\tfrac{x-a}{\vartheta})^{\beta} - e^{\ln(\frac{x-a}{\vartheta})^{\beta}} \right\}$$

make the requisite substitutions,

$$\propto \left(\tfrac{x-a}{\vartheta}\right)^{-1} \exp\left\{ \alpha \tfrac{1}{\sigma\sqrt{\alpha}} \ln(\tfrac{x-a}{\vartheta}) - \alpha e^{\frac{1}{\sigma\sqrt{\alpha}} \ln(\frac{x-a}{\vartheta})} \right\}$$

expand second exponential to second order,

(once more ignoring normalization terms)

$$\propto \left(\tfrac{x-a}{\vartheta}\right)^{-1} \exp\left\{ -\tfrac{1}{2\sigma^2} \left(\ln \tfrac{x-a}{\vartheta}\right)^2 \right\}$$

and reconstitute the normalization constant.

$$= \text{LogNormal}(x \; ; \; a, \vartheta, \sigma)$$

12   Beta Distribution

Beta (β, Beta type I, Pearson type I) distribution [5]:

$$
\begin{aligned}
\text{Beta}(x\ ; a, s, \alpha, \gamma) & \hspace{4cm} (12.1)\\
&= \frac{1}{B(\alpha, \gamma)} \frac{1}{|s|} \left(\frac{x-a}{s}\right)^{\alpha-1} \left(1 - \left(\frac{x-a}{s}\right)\right)^{\gamma-1}\\
&= \text{GenBeta}(x\ ; a, s, \alpha, \gamma, 1)
\end{aligned}
$$

The beta distribution is one member of Person's distribution family, notable for having two roots located at the minimum and maximum of the distribution. The name arises from the beta function in the normalization constant.

Special cases

Special cases of the beta distribution are listed in table 17.1, under $\beta = 1$. With $\alpha < 1$ and $\gamma < 1$ the distribution is U-shaped with a single anti-mode (**U-shaped beta** distribution). If $(\alpha - 1)(\gamma - 1) \leqslant 0$ then the distribution is a monotonic **J-shaped beta** distribution.

Standard beta (Beta) distribution:

$$
\begin{aligned}
\text{StdBeta}(x\ ; \alpha, \gamma) &= \frac{1}{B(\alpha, \gamma)} x^{\alpha-1}(1-x)^{\gamma-1} & (12.2)\\
&= \text{Beta}(x\ ; 0, 1, \alpha, \gamma)\\
&= \text{GenBeta}(x\ ; 0, 1, \alpha, \gamma, 1)
\end{aligned}
$$

The standard beta distribution has two shape parameters, $\alpha > 0$ and $\gamma > 0$, and support $x \in [0, 1]$.

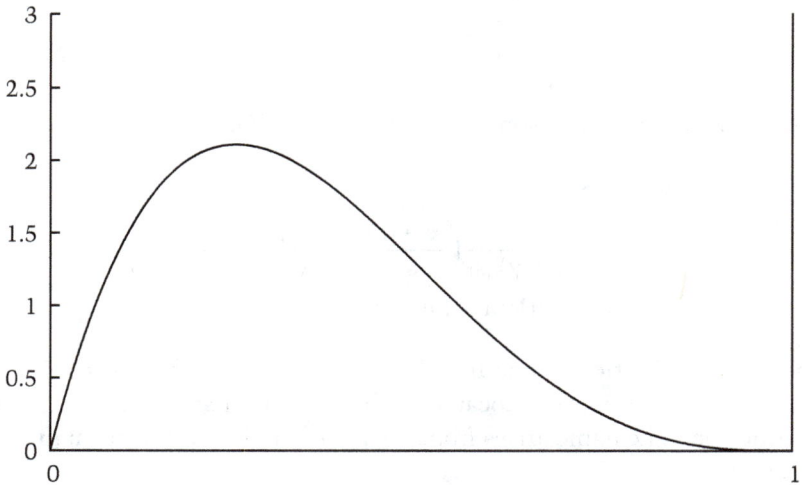

Figure 26: A beta distribution, $\text{Beta}(0, 1, 2, 4)$

Pert (beta-pert) distribution [79, 80] is a subset of the beta distribution, parameterized by minimum (a), maximum (b) and mode (x_{mode}).

$$\text{Pert}(x\,;\,a, b, x_{\text{mode}}) \qquad\qquad (12.3)$$

$$= \frac{1}{B(\alpha, \gamma)(b - a)} \left(\frac{x - a}{b - a}\right)^{\alpha - 1} \left(\frac{b - x}{b - a}\right)^{\gamma - 1}$$

$$x_{\text{mean}} = \frac{a + 4x_{\text{mode}} + b}{6}$$

$$\alpha = \frac{(x_{\text{mean}} - a)(2x_{\text{mode}} - a - b)}{(x_{\text{mode}} - x_{\text{mean}})(b - a)}$$

$$\gamma = \alpha \frac{(b - x_{\text{mean}})}{x_{\text{mean}} - a}$$

$$= \text{Beta}(x\,;\,a, b - a, \alpha, \gamma)$$

$$= \text{GenBeta}(x\,;\,a, b - a, \alpha, \gamma, 1)$$

The PERT (Program Evaluation and Review Technique) distribution is used in project management to estimate task completion times. The **modified pert** distribution replaces the estimate of the mean with $x_{\text{mean}} = \frac{a + \lambda x_{\text{mode}} + b}{2 + \lambda}$, where λ is an additional parameter that controls the spread of the distribu-

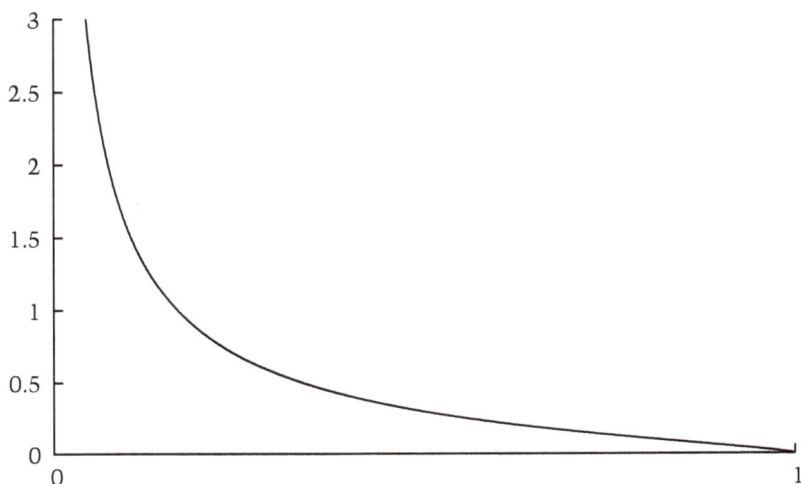

Figure 27: A J-shaped Pearson XII distribution, $\mathrm{Beta}(0, 1, \frac{1}{4}, 1\frac{3}{4})$

tion [80].

Pearson XII distribution [7]:

$$\mathrm{PearsonXII}(x \; ; \; a, b, \alpha) = \frac{1}{B(\alpha, -\alpha + 2)} \frac{1}{|b - a|} \left(\frac{x - a}{b - x} \right)^{\alpha - 1} \qquad (12.4)$$
$$= \mathrm{Beta}(x \; ; \; a, b - a, \alpha, 2 - \alpha)$$
$$= \mathrm{GenBeta}(x \; ; \; a, b - a, \alpha, 2 - \alpha, 1)$$
$$0 < \alpha < 2$$

A monotonic, J-shaped special case of the beta distribution noted by Pearson [7].

Table 12.1: Properties of the beta distribution

Properties

name	$\mathrm{Beta}(x\ ;\ a, s, \alpha, \gamma)$	
PDF	$\dfrac{1}{B(\alpha,\gamma)}\dfrac{1}{\lvert s\rvert}\left(\dfrac{x-a}{s}\right)^{\alpha-1}\left(1-\left(\dfrac{x-a}{s}\right)\right)^{\gamma-1}$	
CDF / CCDF	$\dfrac{B\left(\alpha,\gamma;\frac{x-a}{s}\right)}{B(\alpha,\gamma)} = I(\alpha,\gamma;\frac{x-a}{s})$	$s > 0 \,/\, s < 0$
parameters	$a,\ s,\ \alpha,\ \gamma,\ \text{in } \mathbb{R},$ $\alpha, \gamma \geqslant 0$	
support	$a \geqslant x \geqslant a+s, s > 0 \quad a+s \geqslant x \geqslant a, s < 0$	
mode	$a + s\dfrac{\alpha-1}{\alpha+\gamma-2}$	$\alpha, \gamma > 1$
mean	$a + s\dfrac{\alpha}{\alpha+\gamma}$	
variance	$s^2\dfrac{\alpha\gamma}{(\alpha+\gamma)^2(\alpha+\gamma+1)}$	
skew	$\mathrm{sgn}(s)\dfrac{2(\gamma-\alpha)\sqrt{\alpha+\gamma+1}}{(\alpha+\gamma+2)\sqrt{\alpha\gamma}}$	
ex. kurtosis	$6\dfrac{(\alpha-\gamma)^2(\alpha+\gamma+1)-\alpha\gamma(\alpha+\gamma+2)}{\alpha\gamma(\alpha+\gamma+2)(\alpha+\gamma+3)}$	
entropy	$\ln(\lvert s\rvert) + \ln\big(B(\alpha,\gamma)\big) - (\alpha-1)\psi(\alpha)$ $\quad - (\gamma-1)\psi(\gamma) + (\alpha+\gamma-2)\psi(\alpha+\gamma)$	
MGF	not simple	
CF	${}_1F_1(\alpha;\alpha+\gamma;it)$	

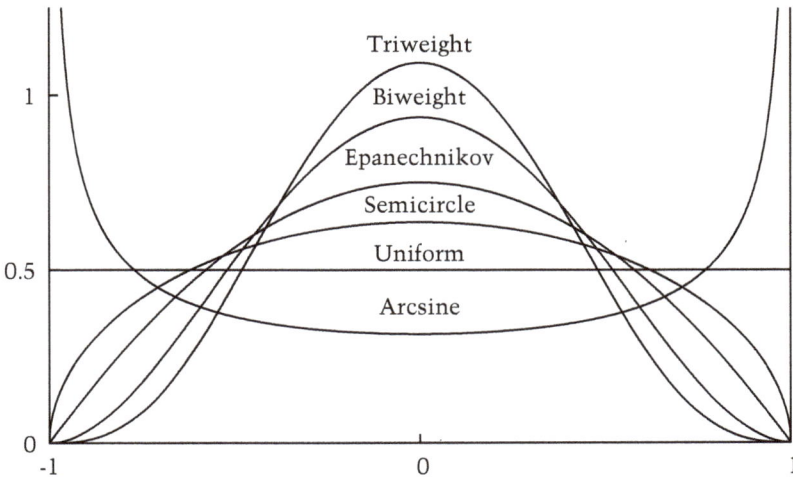

Figure 28: Special cases of the central beta distribution, $\alpha = \frac{1}{2}, 1, \frac{3}{2}, 2, 3, 4$.

Central-beta (Pearson II, symmetric beta, generalized arcsin) distribution [5]:

$$\text{CentralBeta}(x \; ; \; \mu, b, \alpha) = \frac{1}{2^{2\alpha-1}|b|} \frac{\Gamma(2\alpha)}{\Gamma(\alpha)^2} \left(1 - \left(\frac{x-\mu}{b}\right)^2\right)^{\alpha-1} \quad (12.5)$$

$$= \text{Beta}(x \; ; \; \mu - b, 2b, \alpha, \alpha)$$

$$= \text{GenBeta}(x \; ; \; \mu - b, 2b, \alpha, \alpha, 1)$$

A symmetric centered distribution with support $[\mu - b, \mu + b]$.

Arcsine distribution [81]:

$$\text{Arcsine}(x \; ; \; a, s) = \frac{1}{\pi|s|\sqrt{(\frac{x-a}{s})(1 - \frac{x-a}{s})}} \quad (12.6)$$

$$= \text{Beta}(x \; ; \; a, s, \frac{1}{2}, \frac{1}{2})$$

$$= \text{GenBeta}(x \; ; \; a, s, \frac{1}{2}, \frac{1}{2}, 1)$$

Describes the percentage of time spent ahead of the game in a fair coin toss-ing contest [3, 81]. The name comes from the inverse sine function in the cumulative distribution function, $\text{ArcsineCDF}(x \; ; 0, 1) = \frac{2}{\pi} \arcsin(\sqrt{x})$.

Centered arcsine distribution [81]:

$$\text{CenteredArcsine}(x \; ; b) = \frac{1}{2\pi\sqrt{b^2 - x^2}} \tag{12.7}$$
$$= \text{Beta}(x \; ; b, -2b, \tfrac{1}{2}, \tfrac{1}{2})$$
$$= \text{GenBeta}(x \; ; b, -2b, \tfrac{1}{2}, \tfrac{1}{2}, 1)$$

A common variant of the arcsin, with support $x \in [-b, b]$ symmetric about the origin. Describes the position at a random time of a particle engaged in simple harmonic motion with amplitude b [81]. With $b = 1$, the limit-ing distribution of the proportion of time spent on the positive side of the starting position by a simple one dimensional random walk [82].

Semicircle (Wigner semicircle, Sato-Tate) distribution [83]

$$\text{Semicircle}(x \; ; b) = \frac{2}{\pi b^2} \sqrt{b^2 - x^2} \tag{12.8}$$
$$= \text{Beta}(x \; ; -b, 2b, 1\tfrac{1}{2}, 1\tfrac{1}{2})$$
$$= \text{GenBeta}(x \; ; -b, 2b, 1\tfrac{1}{2}, 1\tfrac{1}{2}, 1)$$

As the name suggests, the probability density describes a semicircle, or more properly a half-ellipse. This distribution arises as the distribution of eigenvectors of various large random symmetric matrices.

Epanechnikov (parabolic) distribution [84]:

$$\text{Epanechnikov}(x \; ; \mu, b) = \frac{3}{4} \frac{1}{|b|} \left(1 - \left(\frac{x - \mu}{b} \right)^2 \right) \tag{12.9}$$
$$= \text{CentralBeta}(x \; ; \mu, b, 2)$$
$$= \text{Beta}(x \; ; \mu - b, 2b, 2, 2)$$
$$= \text{GenBeta}(x \; ; \mu - b, 2b, 2, 2, 1)$$

Used in non-parametric kernel density estimation.

Biweight (Quartic) distribution:

$$\text{Biweight}(x\,;\,\mu,b) = \frac{15}{16}\frac{1}{|b|}\left(1-\left(\frac{x-\mu}{b}\right)^2\right)^2 \qquad (12.10)$$

$$= \text{CentralBeta}(x\,;\,\mu,b,3)$$

$$= \text{Beta}(x\,;\,\mu-b,2b,3,3)$$

$$= \text{GenBeta}(x\,;\,\mu-b,2b,3,3,1)$$

Used in non-parametric kernel density estimation.

Triweight distribution:

$$\text{Triweight}(x\,;\,\mu,b) = \frac{35}{32}\frac{1}{|b|}\left(1-\left(\frac{x-\mu}{b}\right)^2\right)^3 \qquad (12.11)$$

$$= \text{CentralBeta}(x\,;\,\mu,b,4)$$

$$= \text{Beta}(x\,;\,\mu-b,2b,4,4)$$

$$= \text{GenBeta}(x\,;\,\mu-b,2b,4,4,1)$$

Used in non-parametric kernel density estimation.

Interrelations

The beta distribution describes the order statistics of a rectangular (1.1) distribution.

$$\text{OrderStatistic}_{\text{Uniform}(a,s)}(x\,;\,\alpha,\gamma) = \text{Beta}(x\,;\,a,s,\alpha,\gamma)$$

Conversely, the uniform (1.1) distribution is a special case of the beta distribution.

$$\text{Beta}(x\,;\,a,s,1,1) = \text{Uniform}(x\,;\,a,s)$$

The beta and gamma distributions are related by

$$\text{StdBeta}(\alpha,\gamma) \sim \frac{\text{StdGamma}_1(\alpha)}{\text{StdGamma}_1(\alpha)+\text{StdGamma}_2(\gamma)}$$

which provides a convenient method of generating beta random variables, given a source of gamma random variables.

The beta distribution is a special case of the generalized beta distribution (17.1), and limits to the gamma distribution (7.1).

$$\text{Gamma}(x \; ; \; a, \theta, \alpha) \; = \; \lim_{\gamma \to \infty} \text{Beta}(x \; ; \; a, \theta\gamma, \alpha, \gamma)$$

The Dirichlet distribution [85, 65] is a multivariate generalization of the beta distribution.

13   Beta Prime Distribution

Beta prime (beta type II, Pearson type VI, inverse beta, variance ratio, gamma ratio, compound gamma, β') distribution [6, 3]:

$$\text{BetaPrime}(x \; ; \; a, s, \alpha, \gamma) \tag{13.1}$$

$$= \frac{1}{B(\alpha, \gamma)} \frac{1}{|s|} \left(\frac{x - a}{s} \right)^{\alpha - 1} \left(1 + \frac{x - a}{s} \right)^{-\alpha - \gamma}$$

$$= \text{GenBetaPrime}(x \; ; \; a, s, \alpha, \gamma, 1)$$

$$\text{for } a, \; s, \; \alpha, \; \gamma \text{ in } \mathbb{R}, \; \alpha > 0, \gamma > 0$$

$$\text{support } x \geqslant a \text{ if } s > 0, \; x \leqslant a \text{ if } s < 0$$

A Pearson distribution (§19) with semi-infinite support, and both roots on the real line. Arises notable as the ratio of gamma distributions, and as the order statistics of the uniform-prime distribution (5.8).

Special cases

Special cases of the beta prime distribution are listed in table 18.1, under $\beta = 1$.

Standard beta prime (beta prime) distribution [6]:

$$\text{StdBetaPrime}(x \; ; \; \alpha, \gamma) = \frac{1}{B(\alpha, \gamma)} x^{\alpha - 1} (1 + x)^{-\alpha - \gamma} \tag{13.2}$$

$$= \text{BetaPrime}(x \; ; \; 0, 1, \alpha, \gamma)$$

$$= \text{GenBetaPrime}(x \; ; \; 0, 1, \alpha, \gamma, 1)$$

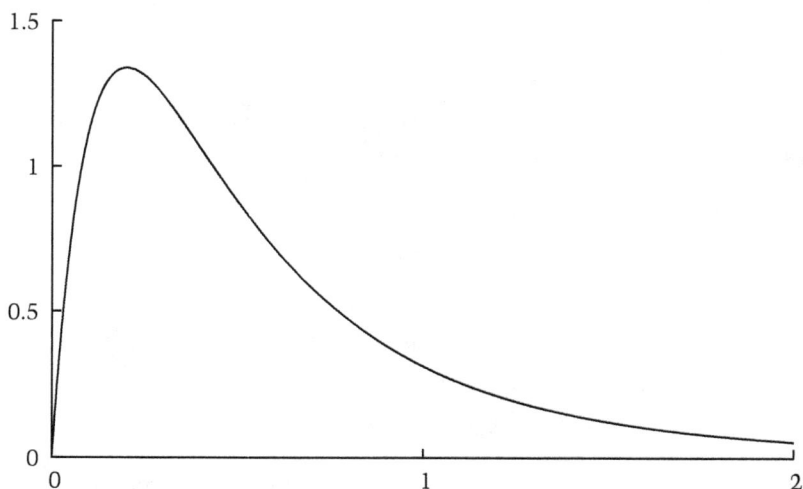

Figure 29: A beta prime distribution, BetaPrime$(0, 1, 2, 4)$

F (Snedecor's F, Fisher-Snedecor, Fisher, Fisher-F, variance-ratio, F-ratio) distribution [86, 87, 3]:

$$F(x \; ; k_1, k_2) = \frac{k_1^{\frac{k_1}{2}} k_2^{\frac{k_2}{2}}}{B(\frac{k_1}{2}, \frac{k_2}{2})} \frac{x^{\frac{k_1}{2}-1}}{(k_2 + k_1 x)^{\frac{1}{2}(k_1+k_2)}} \tag{13.3}$$

$$= \text{BetaPrime}(x \; ; 0, \tfrac{k_2}{k_1}, \tfrac{k_1}{2}, \tfrac{k_2}{2})$$

$$= \text{GenBetaPrime}(x \; ; 0, \tfrac{k_2}{k_1}, \tfrac{k_1}{2}, \tfrac{k_2}{2}, 1)$$

for positive integers $k_1, \; k_2$

An alternative parameterization of the beta prime distribution that derives from the ratio of two chi-squared distributions (7.3) with k_1 and k_2 degrees of freedom.

$$F(k_1, k_2) \sim \frac{\text{ChiSqr}(k_1)/k_1}{\text{ChiSqr}(k_2)/k_2}$$

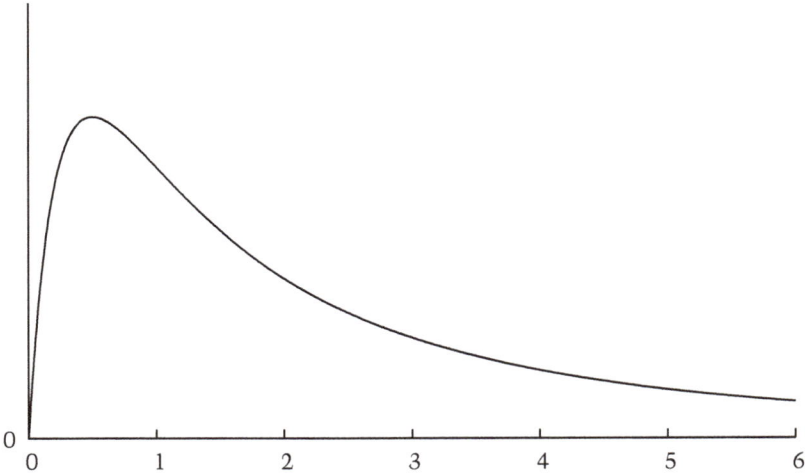

Figure 30: An inverse lomax distribution, InvLomax$(0, 1, 2)$

Inverse Lomax (inverse Pareto) distribution [66]:

$$\text{InvLomax}(x \; ; \; a, s, \alpha) = \frac{\alpha}{|s|} \left(\frac{x-a}{s} \right)^{\alpha-1} \left(1 + \frac{x-a}{s} \right)^{-\alpha-1} \qquad (13.4)$$
$$= \text{BetaPrime}(x \; ; \; a, s, \alpha, 1)$$
$$= \text{GenBetaPrime}(x \; ; \; a, s, \alpha, 1, 1)$$

Interrelations

The standard beta prime distribution is closed under inversion.

$$\text{StdBetaPrime}(\alpha, \gamma) \sim \frac{1}{\text{StdBetaPrime}(\gamma, \alpha)}$$

The beta and beta prime distributions are related by the transformation (§E)

$$\text{StdBetaPrime}(\alpha, \gamma) \sim \left(\frac{1}{\text{StdBeta}(\alpha, \gamma)} - 1 \right)^{-1}$$

Table 13.1: Properties of the beta prime distribution

Properties		
notation	$\text{BetaPrime}(x \; ; a, s, \alpha, \gamma)$	
PDF	$\dfrac{1}{B(\alpha, \gamma)} \dfrac{1}{|s|} \left(\dfrac{x-a}{s} \right)^{\alpha-1} \left(1 + \dfrac{x-a}{s} \right)^{-\alpha-\gamma}$	
CDF / CCDF	$\dfrac{B\left(\alpha, \gamma; (1 + (\frac{x-a}{s})^{-1})^{-1}\right)}{B(\alpha, \gamma)}$	$s > 0 / s < 0$
	$= I\left(\alpha, \gamma; (1 + (\frac{x-a}{s})^{-1})^{-1}\right)$	
parameters	$a, s, \alpha, \gamma,$ in $\mathbb{R}$	
	$\alpha > 0, \gamma > 0$	
support	$x \geqslant a$	$s > 0$
	$x \leqslant a$	$s < 0$
mode	$a + s \dfrac{\alpha - 1}{\gamma + 1}$	$\alpha \geqslant 1$
	a	$\alpha < 1$
mean	$a + s \dfrac{\alpha}{\gamma - 1}$	$\gamma > 1$
variance	$s^2 \dfrac{\alpha(\alpha + \gamma - 1)}{(\gamma - 2)(\gamma - 1)^2}$	$\gamma > 2$
skew	not simple	
ex. kurtosis	not simple	
MGF	none	

and, therefore, the generalized beta prime can be realized as a transformation of the standard beta (12.2) distribution.

$$\text{GenBetaPrime}(a, s, \alpha, \gamma, \beta) \sim a + s\left(\text{StdBeta}(\alpha, \gamma)^{-1} - 1\right)^{-\frac{1}{\beta}}$$

If the scale parameter of a gamma distribution (7.1) is also gamma distributed, the resulting compound distribution is beta prime [88].

$$\text{BetaPrime}(0, s, \alpha, \gamma) \sim \text{Gamma}_2\left(0, \text{Gamma}_1(0, s, \gamma), \alpha\right)$$

The name **compound gamma** distribution is occasionally used for the anchored beta prime distribution (scale parameter, but no location parameter)

The beta prime distribution is a special case of both the generalized beta (17.1) and generalized beta prime (18.1) distributions, and itself limits to the gamma (7.1) and inverse gamma (11.13) distributions.

$$\text{Gamma}(x\,;\,0, \theta, \alpha) = \lim_{\gamma \to \infty} \text{BetaPrime}(x\,;\,0, \theta\gamma, \alpha, \gamma)$$

$$\text{InvGamma}(x\,;\,\theta, \alpha) = \lim_{\gamma \to \infty} \text{BetaPrime}(x\,;\,0, \theta/\gamma, \alpha, \gamma)$$

14   Beta-Exponential Distribution

The **beta-exponential** (Gompertz-Verhulst, generalized Gompertz-Verhulst type III, log-beta, exponential generalized beta type I) distribution [89, 90, 91] is a four parameter, continuous, univariate, unimodal probability density, with semi-infinite support. The functional form in the most straight-forward parameterization is

$$\mathrm{BetaExp}(x \; ; \zeta, \lambda, \alpha, \gamma) = \frac{1}{B(\alpha, \gamma)} \frac{1}{|\lambda|} \, e^{-\alpha \frac{x-\zeta}{\lambda}} \left(1 - e^{-\frac{x-\zeta}{\lambda}}\right)^{\gamma - 1} \tag{14.1}$$

$$\text{for } x, \; \zeta, \; \lambda, \; \alpha, \; \gamma \text{ in } \mathbb{R},$$

$$\alpha, \; \gamma > 0, \quad \tfrac{x-\zeta}{\lambda} > 0 \, .$$

The four real parameters of the beta-exponential distribution consist of a location parameter ζ, a scale parameter λ, and two positive shape parameters α and γ. The **standard beta-exponential** distribution has zero location $\zeta = 0$ and unit scale $\lambda = 1$.

This distribution has a similar shape to the gamma (7.1) distribution. Near the boundary the density scales like $x^{\gamma-1}$, but decays exponentially in the wing.

Special cases

Exponentiated exponential   (generalized exponential, Verhulst) distribution [92, 89, 93]:

$$\mathrm{ExpExp}(x \; ; \zeta, \lambda, \gamma) = \frac{\gamma}{|\lambda|} e^{-\frac{x-\zeta}{\lambda}} \left(1 - e^{-\frac{x-\zeta}{\lambda}}\right)^{\gamma - 1} \tag{14.2}$$

$$= \mathrm{BetaExp}(x \; ; \zeta, \lambda, 1, \gamma)$$

A special case similar in shape to the gamma or Weibull (11.27) distribution. So named because the cumulative distribution function is equal to the exponential distribution function raise to a power.

$$\mathrm{ExpExpCDF}(x \; ; \zeta, \lambda, \gamma) = \left[\, \mathrm{ExpCDF}(x \; ; \zeta, \lambda) \right]^{\gamma}$$

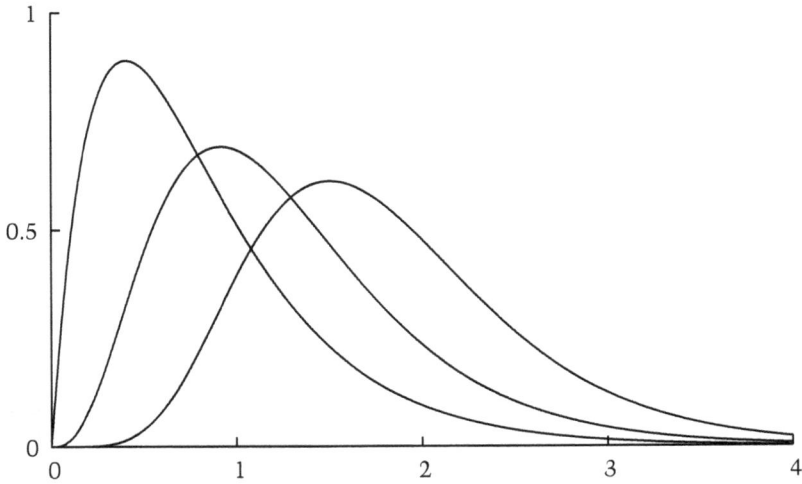

Figure 31: Beta-exponential distributions, (a) BetaExp(x ; 0, 1, 2, 2), (b) BetaExp(x ; 0, 1, 2, 4), (c) BetaExp(x ; 0, 1, 2, 8).

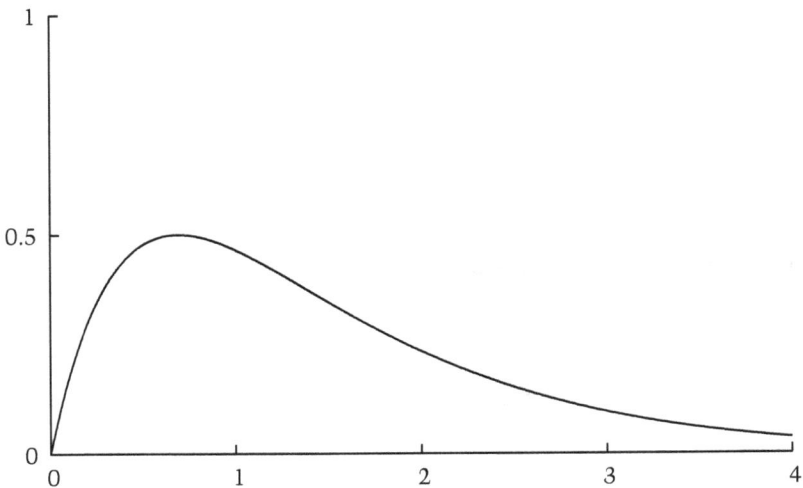

Figure 32: Exponentiated exponential distribution, ExpExp(x ; 0, 1, 2).

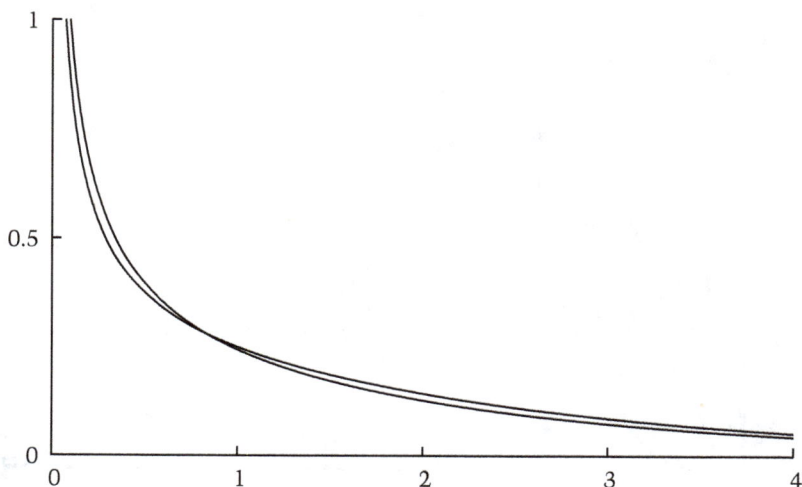

Figure 33: Hyperbolic sine HyperbolicSine(x ; $\frac{1}{2}$) and Nadarajah-Kotz NadarajahKotz(x) distributions.

Hyperbolic sine distribution [1]:

$$\text{HyperbolicSine}(x \; ; \zeta, \lambda, \gamma) = \frac{1}{B(\frac{1-\gamma}{2}, \gamma)} \frac{1}{|\lambda|} \left(e^{+\frac{x-\zeta}{2\lambda}} - e^{-\frac{x-\zeta}{2\lambda}} \right)^{\gamma-1} \quad (14.3)$$

$$= \frac{2^{\gamma-1}}{B(\frac{1-\gamma}{2}, \gamma)|\lambda|} \left(\sinh(\tfrac{x-\zeta}{2\lambda}) \right)^{\gamma-1}$$

$$= \text{BetaExp}(x \; ; \zeta, \lambda, \tfrac{1-\gamma}{2}, \gamma), \quad 0 < \gamma < 1$$

Compare to the hyperbolic secant distribution (15.6).

Nadarajah-Kotz distribution [90, 1] :

$$\text{NadarajahKotz}(x \; ; \zeta, \lambda) = \frac{1}{\pi|\lambda|} \frac{1}{\sqrt{e^{\frac{x-\zeta}{\lambda}} - 1}} \quad (14.4)$$

$$= \text{BetaExp}(x \; ; \zeta, \lambda, \tfrac{1}{2}, \tfrac{1}{2})$$

A notable special case when $\alpha = \gamma = \frac{1}{2}$. The cumulative distribution

Table 14.1: Special cases of the beta-exponential family

(14.1)	beta-exponential	ζ	λ	α	γ	
	std. beta-exponential	0	1	.	.	
(14.2)	exponentiated exponential	.	.	1	.	
(14.3)	hyperbolic sine	.	.	$\frac{1}{2}(1\text{-}\gamma)$	γ	$0 < \gamma < 1$
(14.4)	Nadarajah-Kotz	.	.	$\frac{1}{2}$	$\frac{1}{2}$	
(2.1)	exponential	.	.	1	1	

function has the simple form

$$\text{NadarajahKotzCDF}(x\;;\,0,1) = \frac{2}{\pi}\arctan\sqrt{\exp(x)-1}\,.$$

Interrelations

The beta-exponential distribution is a limit of the generalized beta distribution (§12). The analogous limit of the generalized beta prime distribution (§13) results in the beta-logistic family of distributions (§15).

The beta-exponential distribution is the log transform of the beta distribution (12.1).

$$\text{StdBetaExp}(\alpha,\gamma) \sim -\ln\big(\text{StdBeta}(\alpha,\gamma)\big)$$

It follows that beta-exponential variates are related to ratios of gamma variates.

$$\text{StdBetaExp}(\alpha,\gamma) \sim -\ln\frac{\text{StdGamma}_1(\alpha)}{\text{StdGamma}_1(\alpha)+\text{StdGamma}_2(\gamma)}$$

The beta-exponential distribution describes the order statistics (§C) of the exponential distribution (2.1).

$$\text{OrderStatistic}_{\text{Exp}(\zeta,\lambda)}(x\;;\,\gamma,\alpha) = \text{BetaExp}(x\;;\,\zeta,\lambda,\alpha,\gamma)$$

With $\gamma = 1$ we recover the exponential distribution.

$$\text{BetaExp}(x\;;\,\zeta,\lambda,\alpha,1) = \text{Exp}(x\;;\,\zeta,\tfrac{\lambda}{\alpha})$$

Table 14.2: Properties of the beta-exponential distribution

Properties

notation	$\text{BetaExp}(x \; ; \zeta, \lambda, \alpha, \gamma)$	
PDF	$\dfrac{1}{B(\alpha,\gamma)} \dfrac{1}{|\lambda|} e^{-\alpha \frac{x-\zeta}{\lambda}} \left(1 - e^{-\frac{x-\zeta}{\lambda}}\right)^{\gamma-1}$	
CDF/CCDF	$I\left(\alpha, \gamma; e^{-\frac{x-\zeta}{\lambda}}\right)$	$\lambda > 0 \, / \, \lambda < 0$
parameters	$\zeta, \lambda, \alpha, \gamma$ in $\mathbb{R}$	
	$\alpha, \gamma > 0$	
support	$x \geqslant \zeta$	$\lambda > 0$
	$x \leqslant \zeta$	$\lambda < 0$
mean	$\zeta + \lambda[\psi(\alpha+\gamma) - \psi(\alpha)]$	[90]
variance	$\lambda^2[\psi_1(\alpha) - \psi_1(\alpha+\gamma)]$	[90]
skew	$-\,\text{sgn}(\lambda)\,\left[\psi_2(\alpha) - \psi_2(\alpha+\gamma)\right]$	
	$\Big/ \left[\psi_1(\alpha) - \psi_1(\alpha+\gamma)\right]^{\frac{3}{2}}$	[90]
ex. kurtosis	$\left[3\psi_1(\alpha)^2 - 6\psi_1(\alpha)\psi_1(\alpha+\gamma) + 3\psi_1(\alpha+\gamma)^2 + \psi_3(\alpha)\right.$	
	$\left. - \psi_3(\alpha+\gamma)\right] \Big/ \left[\psi_1(\alpha) - \psi_1(\alpha+\gamma)\right]^2$	[90]
entropy	$\ln|\lambda| + \ln B(\alpha,\gamma) + (\alpha+\gamma-1)\psi(\alpha+\gamma)$	
	$- (\gamma-1)\psi(\gamma) - \alpha\psi(\alpha)$	[90]
MGF	$e^{\zeta t} \dfrac{B(\alpha - \lambda t, \gamma)}{B(\alpha,\gamma)}$	[90]
CF	$e^{i\zeta t} \dfrac{B(\alpha - i\lambda t, \gamma)}{B(\alpha,\gamma)}$	[90]

The beta-exponential distribution is a limit of the generalized beta distribution (17.1), and itself limits to the gamma-exponential distriution (8.1).

$$\text{GammaExp}(x \; ; \nu, \lambda, \alpha) = \lim_{\gamma \to \infty} \text{BetaExp}(x \; ; \nu + \lambda/\ln\gamma, \lambda, \alpha, \gamma)$$

15 Beta-Logistic Distribution

The **beta-logistic** (Prentice, beta-prime exponential, generalized logistic type IV, exponential generalized beta prime, exponential generalized beta type II, log-F, generalized F, Fisher-z, generalized Gompertz-Verhulst type II) distribution [94, 95, 3, 96] is a four parameter, continuous, univariate, unimodal probability density, with infinite support. The functional form in the most straightforward parameterization is

$$\text{BetaLogistic}(x \; ; \zeta, \lambda, \alpha, \gamma) = \frac{1}{B(\alpha, \gamma)|\lambda|} \frac{e^{-\alpha \frac{x-\zeta}{\lambda}}}{\left(1 + e^{-\frac{x-\zeta}{\lambda}}\right)^{\alpha+\gamma}}$$

$$x, \zeta, \lambda, \alpha, \gamma \text{ in } \mathbb{R} \qquad (15.1)$$

$$\alpha, \gamma > 0$$

The four real parameters consist of a location parameter ζ, a scale parameter λ, and two positive shape parameters α and γ. The **standard beta-logistic** distribution has zero location $\zeta = 0$ and unit scale $\lambda = 1$.

The beta-logistic distribution is perhaps most commonly referred to as 'generalized logistic', but this terminology is ambiguous, since many types of generalized logistic distribution have been investigated, and this distribution is not 'generalized' in the same sense used elsewhere in this survey (See 'generalized' §A). Therefore, we select the name 'beta-logistic' as a less ambiguous terminology that mirrors the names beta, beta-prime, and beta-exponential.

Special cases

Burr type II (generalized logistic type I, exponential-Burr, skew-logistic) distribution [97, 2]:

$$\text{BurrII}(x \; ; \zeta, \lambda, \gamma) = \frac{\gamma}{|\lambda|} \frac{e^{-\frac{x-\zeta}{\lambda}}}{\left(1 + e^{-\frac{x-\zeta}{\lambda}}\right)^{\gamma+1}} \qquad (15.2)$$

$$= \text{BetaLogistic}(x \; ; \zeta, \lambda, 1, \gamma)$$

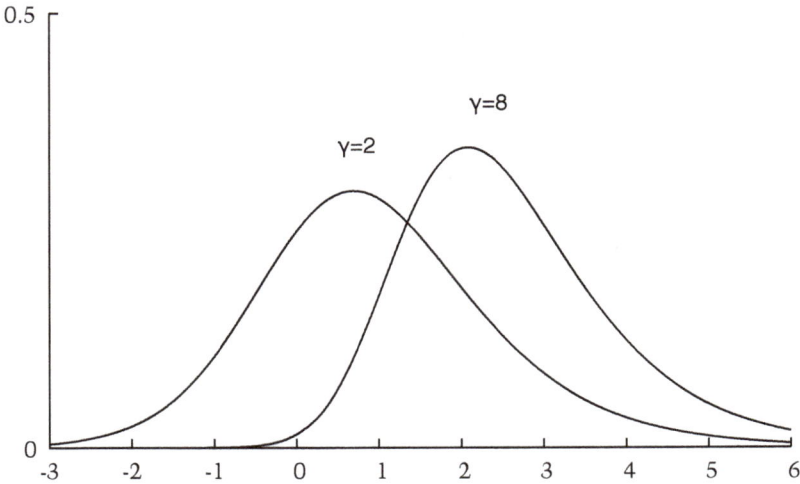

Figure 34: Burr type II distributions, $\text{BurrII}(x \; ; 0, 1, \gamma)$

Reversed Burr type II (generalized logistic type II) distribution [2]:

$$\text{RevBurrII}(x \; ; \alpha) = \frac{\gamma}{|\lambda|} \frac{e^{+\frac{x-\zeta}{\lambda}}}{\left(1 + e^{+\frac{x-\zeta}{\lambda}}\right)^{\gamma+1}} \tag{15.3}$$

$$= \text{BurrII}(x \; ; \zeta, -\lambda, \gamma)$$
$$= \text{BetaLogistic}(x \; ; \zeta, -\lambda, 1, \gamma)$$
$$= \text{BetaLogistic}(x \; ; \zeta, +\lambda, \gamma, 1)$$

By setting the λ parameter to 1 (instead of α) we get a reversed Burr type II.

Table 15.1: Special cases of the beta-logistic distribution

		ζ	λ	α	γ
(15.1)	Beta-Logistic				
(15.2)	Burr type II	.	.	1	.
(15.3)	Reversed Burr type II	.	.	.	1
(15.4)	Central Logistic	.	.	α	α
(15.5)	Logistic	.	.	1	1
(15.6)	Hyperbolic secant	.	.	$\frac{1}{2}$	$\frac{1}{2}$

Table 15.2: Properties of the beta-logistic distribution

Properties	
notation	$\text{BetaLogistic}(x\ ;\ \zeta,\lambda,\alpha,\gamma)$
PDF	$\dfrac{1}{B(\alpha,\gamma)|\lambda|}\ \dfrac{e^{-\alpha\frac{x-\zeta}{\lambda}}}{\left(1+e^{-\frac{x-\zeta}{\lambda}}\right)^{\alpha+\gamma}}$
CDF / CCDF	$\dfrac{B\left(\gamma,\alpha;(1+e^{-\frac{x-\zeta}{\lambda}})^{-1}\right)}{B(\alpha,\gamma)}$ $\quad\quad \lambda>0/\lambda<0\,[1]$ $\qquad\qquad = I\left(\gamma,\alpha;(1+e^{-\frac{x-\zeta}{\lambda}})^{-1}\right)$
parameters	$\zeta,\ \lambda,\ \alpha,\gamma$ in $\mathbb{R}$ $\alpha,\ \gamma>0$
support	$x\in[-\infty,+\infty]$
mean	$\zeta+\lambda[\psi(\gamma)-\psi(\alpha)]$
variance	$\lambda^2[\psi_1(\alpha)+\psi_1(\gamma)]$
skew	$\text{sgn}(\lambda)\ \dfrac{\psi_2(\gamma)-\psi_2(\alpha)}{[\psi_1(\alpha)+\psi_1(\gamma)]^{3/2}}$
ex. kurtosis	$\dfrac{\psi_3(\alpha)+\psi_3(\gamma)}{[\psi_1(\alpha)+\psi_1(\gamma)]^2}$
MGF	$e^{\zeta t}\dfrac{\Gamma(\alpha-\lambda t)\Gamma(\gamma+\lambda t)}{\Gamma(\alpha)\Gamma(\gamma)}$ $\qquad\qquad$ [3]
CF	$e^{i\zeta t}\dfrac{\Gamma(\alpha+i\lambda t)\Gamma(\gamma-i\lambda t)}{\Gamma(\alpha)\Gamma(\gamma)}$

Central-logistic (generalized logistic type III, symmetric Prentice, symmetric beta-logistic) distribution [3]:

$$\text{CentralLogistic}(x \; ; \; \zeta, \lambda, \alpha) = \frac{1}{B(\alpha, \alpha)|\lambda|} \frac{e^{-\alpha \frac{x-\zeta}{\lambda}}}{\left(1 + e^{-\frac{x-\zeta}{\lambda}}\right)^{2\alpha}} \tag{15.4}$$

$$= \frac{1}{B(\alpha, \alpha)|\lambda|} \left[\tfrac{1}{2} \text{sech}\left(\tfrac{x-\zeta}{2\lambda}\right)\right]^{2\alpha}$$

$$= \text{BetaLogistic}(x \; ; \; \zeta, \lambda, \alpha, \alpha)$$

With equal shape parameters the beta-logistic is symmetric. This distribution limits to the Laplace distribution (3.1).

Logistic (sech-square, hyperbolic secant square, logit) distribution [98, 99, 3]:

$$\text{Logistic}(x \; ; \; \zeta, \lambda) = \frac{1}{|\lambda|} \frac{e^{-\frac{x-\zeta}{\lambda}}}{\left(1 + e^{-\frac{x-\zeta}{\lambda}}\right)^2} \tag{15.5}$$

$$= \frac{1}{4|\lambda|} \text{sech}^2\left(\frac{x-\zeta}{\lambda}\right)$$

$$= \text{BetaLogistic}(x \; ; \; \zeta, \lambda, 1, 1)$$

Hyperbolic secant (inverse hyperbolic cosine, inverse cosh) distribution [3, 100, 101]:

$$\text{HyperbolicSecant}(x \; ; \; \zeta, \lambda) = \frac{1}{\pi|\lambda|} \frac{1}{e^{+\frac{x-\zeta}{2\lambda}} + e^{-\frac{x-\zeta}{2\lambda}}} \tag{15.6}$$

$$= \frac{1}{2\pi|\lambda|} \text{sech}\left(\tfrac{x-\zeta}{2\lambda}\right)$$

$$= \text{BetaLogistic}(x \; ; \; \zeta, \lambda, \tfrac{1}{2}, \tfrac{1}{2})$$

The hyperbolic secant cumulative distribution function features the Gudermannian sigmoidal function, $\text{gd}(z)$.

$$\text{HyperbolicSecantCDF}(x \; ; \; \zeta, \lambda) = \frac{1}{\pi} \text{gd}(\frac{x-\zeta}{2\lambda})$$

$$= \frac{2}{\pi} \arctan(e^{\frac{x-\zeta}{2\lambda}}) - \frac{1}{2}$$

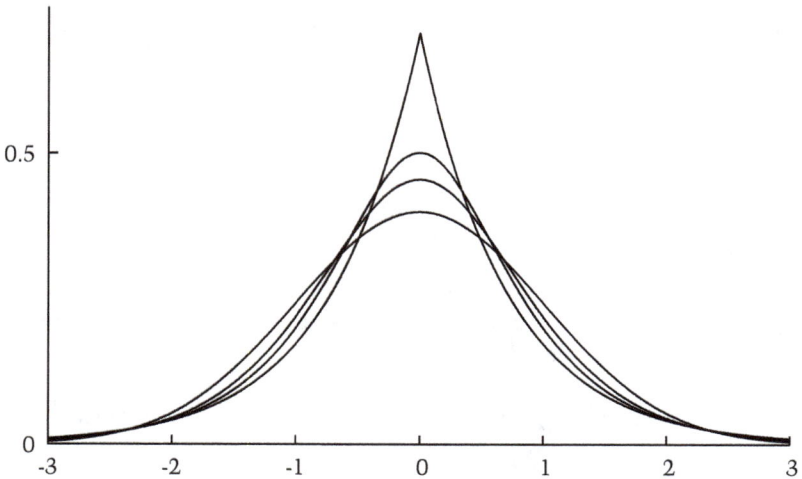

Figure 35: Special cases of the symmetric central-logistic distribution (15.4): Standardized (zero mean, unit variance) normal ($\alpha \to \infty$), logistic ($\alpha = 1$), hyperbolic secant ($\alpha = \frac{1}{2}$), and Laplace ($\alpha \to 0$) (low to high peaks).

The standardized hyperbolic secant distribution (zero mean, unit variance) is HyperbolicSecant$(x\ ;\ 0, 1/\pi)$.

Interrelations

The beta-logistic distribution arises as a limit of the generalized beta-prime distribution (§13). The analogous limit of the generalized beta distribution leads to the beta-exponential family (§14).

The beta-logistic distribution is the log transform of the beta prime distribution.

$$\text{BetaLogistic}(0, 1, \alpha, \gamma) \sim -\ln \text{BetaPrime}(0, 1, \alpha, \gamma)$$

It follows that beta-logistic variates are related to ratios of gamma variates.

$$\text{BetaLogistic}(\zeta, \lambda, \alpha, \gamma) \sim \zeta - \lambda \ln \frac{\text{StdGamma}_1(\gamma)}{\text{StdGamma}_2(\alpha)}$$

Negating the scale parameter is equivalent to interchanging the two shape parameters.

$$\text{BetaLogistic}(x \; ; \; \zeta, +\lambda, \alpha, \gamma) = \text{BetaLogistic}(x \; ; \; \zeta, -\lambda, \gamma, \alpha)$$

The beta-logistic distribution, with integer α and γ is the logistic order statistics distribution [102, 20] (§C).

$$\text{OrderStatistic}_{\text{Logistic}(\zeta,\lambda)}(x \; ; \; \gamma, \alpha) = \text{BetaLogistic}(x \; ; \; \zeta, \lambda, \alpha, \gamma)$$

The beta-logistic limits to the gamma exponential (8.1) and Laplace (3.1) distributions.

$$\text{GammaExp}(x \; ; \; \nu, \lambda, \alpha) = \lim_{\gamma \to \infty} \text{BetaLogistic}(x \; ; \; \nu + \lambda/\ln\gamma, \lambda, \alpha, \gamma)$$

$$\text{Laplace}(x \; ; \; \eta, \theta) = \lim_{\alpha \to 0} \text{BetaLogistic}(x \; ; \; \eta, \theta\alpha\,\alpha, \alpha)$$

16   Pearson IV Distribution

Pearson IV (skew-t) distribution [5, 103] is a four parameter, continuous, univariate, unimodal probability density, with infinite support. The functional form is

$$\text{PearsonIV}(x \; ; \; a, s, m, v) \tag{16.1}$$

$$= \frac{{}_2F_1(-iv, iv; m; 1)}{|s|B(m - \frac{1}{2}, \frac{1}{2})} \left(1 + \left(\frac{x-a}{s}\right)^2\right)^{-m} \exp\left\{-2v \; \arctan\left(\frac{x-a}{s}\right)\right\}$$

$$= \frac{{}_2F_1(-iv, iv; m; 1)}{|s|B(m - \frac{1}{2}, \frac{1}{2})} \left(1 + i\frac{x-a}{s}\right)^{-m+iv} \left(1 - i\frac{x-a}{s}\right)^{-m-iv}$$

$$x, a, s, m, v \in \mathbb{R}$$

$$m > \frac{1}{2}$$

Note that the two forms are equivalent, since $\arctan(z) = \frac{1}{2}i \ln \frac{1-iz}{1+iz}$. The first form is more conventional, but the second form displays the essential simplicity of this distribution. The density is an analytic function with two singularities, located at conjugate points in the complex plain, with conjugate, complex order. This is the one member of the Pearson distribution family that has not found significant utility.

Interrelations

The distribution parameters obey the symmetry

$$\text{PearsonIV}(x \; ; \; a, s, m, v) = \text{PearsonIV}(x \; ; \; a, -s, m, -v).$$

Setting the complex part of the exponents to zero, $v = 0$, gives the Pearson VII family (9.1), which includes the Cauchy and Student's t distributions.

$$\text{PearsonIV}(x \; ; \; a, s, m, 0) = \text{PearsonVII}(x \; ; \; a, s, m)$$

Suitable rescaled, the exponentiated arctan limits to an exponential of the reciprocal argument.

$$\lim_{\nu\to\infty} \exp(-2\nu \arctan(-2\nu x) - \pi\nu) = e^{-\frac{1}{x}}$$

Consequently, the high ν limit of the Pearson IV distribution is an inverse gamma (Pearson V) distribution (11.13), which acts an intermediate distribution between the beta prime (Pearson VI) and Pearson IV distributions.

$$\lim_{\nu\to\infty} \text{PearsonIV}(x \; ; \; 0, -\tfrac{\theta}{2\nu}, \tfrac{\alpha+1}{2}, \nu) = \text{InvGamma}(x \; ; \; \theta, \alpha)$$

The inverse exponential distribution (11.14) is therefore also a special case when $\alpha = 1$ ($m = 1$).

Table 16.1: Properties of the Pearson IV distribution

Properties

notation	$\mathrm{PearsonIV}(x \; ; a, s, m, v)$
PDF	$\dfrac{{}_2F_1(-iv, iv; m; 1)}{\|s\|B(m - \frac{1}{2}, \frac{1}{2})}\left(1 + \left(\dfrac{x - a}{s}\right)^2\right)^{-m}$
	$\times \exp\left\{-2v \arctan\left(\dfrac{x - a}{s}\right)\right\}$
CDF	$\mathrm{PearsonIV}(x \; ; a, s, m, v)$
	$\times \dfrac{\|s\|}{2m - 1}\left(i - \dfrac{x - a}{s}\right){}_2F_1\left(1, m + iv; 2m; \dfrac{2}{i - i\frac{x-a}{s}}\right)$
parameters	$a, \; s, \; m, \; v$ in $\mathbb{R}$
	$m > \frac{1}{2}$
support	$x \in [-\infty, +\infty]$
mode	$a - \dfrac{sv}{m}$
mean	$a - \dfrac{sv}{(m - 1)} \qquad (m > 1)$
variance	$\dfrac{s^2}{2m - 3}(1 + \dfrac{v^2}{(m - 1)^2}) \qquad (m > \frac{3}{2})$
skew	not simple
ex. kurtosis	not simple

17   Generalized Beta Distribution

The **generalized beta** (beta-power) distribution [51] is a five parameter, continuous, univariate, unimodal probability density, with finite or semi infinite support. The functional form in the most straightforward parameterizaton is

$$\text{GenBeta}(x \; ; a, s, \alpha, \gamma, \beta) \hfill (17.1)$$

$$= \frac{1}{B(\alpha, \gamma)} \left| \frac{\beta}{s} \right| \left(\frac{x-a}{s} \right)^{\alpha\beta - 1} \left(1 - \left(\frac{x-a}{s} \right)^{\beta} \right)^{\gamma - 1}$$

$$\text{for } x, a, \theta, \alpha, \gamma, \beta \text{ in } \mathbb{R},$$

$$\alpha > 0, \; \gamma > 0$$

$$\text{support } x \in [a, a+s], s > 0, \; \beta > 0$$

$$x \in [a+s, a], s < 0, \; \beta > 0$$

$$x \in [a+s, +\infty], s > 0, \; \beta < 0$$

$$x \in [-\infty, a+s], s < 0, \; \beta < 0$$

The generalized beta distribution arises as the Weibullization of the standard beta distribution, $x \to (\frac{x-a}{s})^{\beta}$, and as the order statistics of the power function distribution (5.1). The parameters consist of a location parameter a, shape parameter s, Weibull power parameter β, and two shape parameters α and γ.

Special Cases

The beta distribution (β=1) and specializations are described in (§12).

Kumaraswamy (minimax) distribution [104, 8, 105]:

$$\text{Kumaraswamy}(x \; ; a, s, \gamma, \beta) = \gamma \left| \frac{\beta}{s} \right| \left(\frac{x-a}{s} \right)^{\beta - 1} \left(1 - \left(\frac{x-a}{s} \right)^{\beta} \right)^{\gamma - 1}$$

$$\hfill (17.2)$$

$$= \text{GenBeta}(x \; ; a, s, 1, \gamma, \beta)$$

Proposed as an alternative to the beta distribution for modeling bounded variables, since the cumulative distribution function has a simple closed

Table 17.1: Special cases of generalized beta

(17.1)	generalized beta	a	s	α	γ	β	
(17.2)	Kumaraswamy	.	.	1	.	.	
(12.1)	beta	.	.	.	.	1	
(12.2)	standard beta	0	1	.	.	1	
(12.1)	beta, U shaped	.	.	<1	<1	1	
(12.1)	beta, J shaped	.	.	.	.	1	$(\alpha\text{-}1)(\gamma\text{-}1) \leqslant 0$
(12.5)	central beta	.	.	α	α	1	
(12.6)	arcsine	.	.	$\frac{1}{2}$	$\frac{1}{2}$	1	
(12.8)	semicircle	-b	2b	$1\frac{1}{2}$	$1\frac{1}{2}$	1	
(12.9)	Epanechnikov	.	.	2	2	1	
(12.10)	biweight	.	.	3	3	1	
(12.11)	triweight	.	.	4	4	1	
(12.4)	Pearson XII	.	.	.	2-α	1	$\alpha < 2$
(13.1)	beta-prime	.	.	.	.	-1	
(5.1)	power function	.	.	1	1	.	
(1.1)	uniform	.	.	1	1	1	
(1.1)	standard uniform	0	1	1	1	1	

Table 17.2: Properties of the generalized beta distribution

Properties		
name	$\text{GenBeta}(x \ ; \ a, s, \alpha, \gamma, \beta)$	
PDF	$\dfrac{1}{B(\alpha,\gamma)}\left\lvert\dfrac{\beta}{s}\right\rvert \left(\dfrac{x-a}{s}\right)^{\alpha\beta-1}\left(1-\left(\dfrac{x-a}{s}\right)^{\beta}\right)^{\gamma-1}$	
CDF / CCDF	$\dfrac{B\left(\alpha,\gamma;\,(\frac{x-a}{s})^{\beta}\right)}{B(\alpha,\gamma)}$	$\frac{\beta}{s} > 0 \,/\, \frac{\beta}{s} < 0$
	$= I\left(\alpha,\gamma;\,(\frac{x-a}{s})^{\beta}\right)$	
parameters	$a, s, \alpha, \gamma, \beta, \text{ in } \mathbb{R},$	
	$\alpha, \gamma \geqslant 0$	
support	$x \in [a, a+s],$	$0 < s, \ 0 < \beta$
	$x \in [a+s, a],$	$s < 0, \ 0 < \beta$
	$x \in [a+s, +\infty],$	$0 < s, \ \beta < 0$
	$x \in [-\infty, a+s],$	$s < 0, \ \beta < 0$
mean	$a + \dfrac{sB(\alpha+\frac{1}{\beta},\gamma)}{B(\alpha,\gamma)}$	$\alpha + \frac{1}{\beta} > 0$
variance	$\dfrac{s^2 B(\alpha+\frac{2}{\beta},\gamma)}{B(\alpha,\gamma)} - \dfrac{s^2 B(\alpha+\frac{1}{\beta},\gamma)^2}{B(\alpha,\gamma)^2}$	
skew	not simple	
ex. kurtosis	not simple	
MGF	none	
$E(X^h)$	$\dfrac{s^h B(\alpha+\frac{h}{\beta},\gamma)}{B(\alpha,\gamma)}$	$a = 0, \ \alpha + \frac{h}{\beta} > 0 \ [51]$

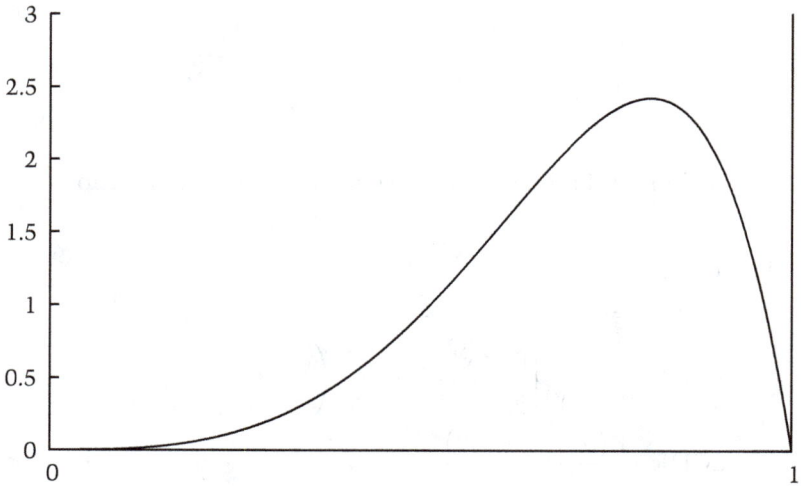

Figure 36: A Kumaraswamy distribution, Kumaraswamy$(0, 1, 2, 4)$

form,

$$\text{KumaraswamyCDF}(x\ ;\ 0, 1, \gamma, \beta) = 1 - (1 - x^\beta)^\gamma.$$

Interrelations

The generalized beta distribution describes the order statistics of a power function distribution (5.1).

$$\text{OrderStatistic}_{\text{PowerFn}(a,s,\beta)}(x\ ;\ \alpha, \gamma) = \text{GenBeta}(x\ ;\ a, s, \alpha, \gamma, \beta)$$

Conversely, the power function (5.1) distribution is a special case of the generalized beta distribution.

$$\text{GenBeta}(x\ ;\ a, s, 1, 1, \beta) = \text{PowerFn}(x\ ;\ a, s, \beta)$$

Setting $\beta = 1$ yields the beta distribution (12.1),

$$\text{GenBeta}(x\ ;\ a, s, \alpha, \gamma, 1) = \text{Beta}(x\ ;\ a, s, \alpha, \gamma)\ ,$$

and setting $\beta = -1$ yields the beta prime (or inverse beta) distribution (13.1),

$$\mathrm{GenBeta}(x \; ; \; a, s, \alpha, \gamma, -1) = \mathrm{BetaPrime}(x \; ; \; a + s, s, \gamma, \alpha) \; .$$

The beta (§12) and beta prime (§13) distributions have many named special cases, see tables 17.1 and 18.1.

The unit gamma distribution (10.1) arises in the limit $\lim_{\beta \to 0}$ with $\alpha\beta =$ constant,

$$\lim_{\beta \to 0} \mathrm{GenBeta}(x \; ; \; a, s, \tfrac{\delta}{\beta}, \gamma, \beta) = \mathrm{UnitGamma}(x \; ; \; a, s, \gamma, \delta) \; .$$

In the limit $\gamma \to \infty$ (or equivalently $\alpha \to \infty$) we obtain the Amoroso distribution (11.1) with semi-infinite support, the parent of the gamma distribution family [51],

$$\lim_{\gamma \to \infty} \mathrm{GenBeta}(x \; ; \; a, \theta\gamma^{\frac{1}{\beta}}, \alpha, \gamma, \beta) = \mathrm{Amoroso}(x \; ; \; a, \theta, \alpha, \beta) \; .$$

The limit $\lim_{\beta \to +\infty}$ yields the beta-exponential distribution (14.1)

$$\lim_{\beta \to +\infty} \mathrm{GenBeta}(x \; ; \; \zeta + \beta\lambda, -\beta\lambda, \alpha, \gamma, \beta) = \mathrm{BetaExp}(x \; ; \; \zeta, \lambda, \alpha, \gamma) \; .$$

18 Generalized Beta Prime Distribution

The **generalized beta-prime** (Feller-Pareto, beta-log-logistic, generalized gamma ratio, Majumder-Chakravart, generalized beta type II, generalized Feller-Pareto) distribution [67, 51, 106] is a five parameter, continuous, univariate, unimodal probability density, with semi-infinite support. The functional form in the most straightforward parameterization is

$$
\text{GenBetaPrime}(x \; ; a, s, \alpha, \gamma, \beta) \tag{18.1}
$$

$$
= \frac{1}{B(\alpha, \gamma)} \left| \frac{\beta}{s} \right| \left(\frac{x-a}{s} \right)^{\alpha\beta-1} \left(1 + \left(\frac{x-a}{s} \right)^{\beta} \right)^{-\alpha-\gamma}
$$

$$
a, \; s, \; \alpha, \; \gamma, \; \beta \text{ in } \mathbb{R}, \quad \alpha, \; \gamma > 0
$$

The five real parameters of the generalized beta prime distribution consist of a location parameter a, scale parameter s, two shape parameters, α and γ, and the Weibull power parameter β. The shape parameters, α and γ, are positive.

The generalized beta prime arises as the Weibull transform of the standard beta prime distribution (13.2), and as order statistics of the log-logistic distribution. The Amoroso distribution is a limiting form, and a variety of other distributions occur as special cases. (See Table 18.1). These distributions are most often encountered as parametric models for survival statistics developed by economists and actuaries.

Special cases

Transformed beta distribution [51, 107]:

$$
\text{TransformedBeta}(x \; ; s, \alpha, \gamma, \beta) \tag{18.2}
$$

$$
= \frac{1}{B(\alpha, \gamma)} \left| \frac{\beta}{s} \right| \left(\frac{x}{s} \right)^{\alpha\beta-1} \left(1 + \left(\frac{x}{s} \right)^{\beta} \right)^{-\alpha-\gamma}
$$

$$
= \text{GenBetaPrime}(x \; ; 0, s, \alpha, \gamma, \beta)
$$

A generalized beta prime distribution without a location parameter, $a = 0$.

Burr (Burr type XII, Pareto type IV, beta-P, Singh-Maddala, generalized log-

Table 18.1: Special cases of generalized beta prime

(18.1)	generalized beta prime	a	s	α	γ	β
(18.3)	Burr	.	.	1	.	.
(18.4)	Dagum	0	1	.	1	.
(18.5)	paralogistic	0	1	1	β	.
(18.6)	inverse paralogistic	0	1	β	1	.
(18.7)	log-logistic	0	.	1	1	.
(18.1)	transformed beta	0	.	.	.	.
(18.10)	half gen. Pearson VII	.	.	$\frac{1}{\beta}$	$m-\frac{1}{\beta}$	.
(13.1)	beta prime	.	.	.	.	1
(5.6)	Lomax	.	.	1	.	1
(13.4)	inverse Lomax	.	.	.	1	1
(13.2)	std. beta-prime	0	1	.	.	1
(13.3)	F	0	$\frac{k_2}{k_1}$	$\frac{k_1}{2}$	$\frac{k_2}{2}$	1
(5.8)	uniform-prime	.	.	1	1	1
(5.7)	exponential ratio	0	.	1	1	1
(18.8)	half-Pearson VII	.	.	$\frac{1}{2}$	.	2
(18.9)	half-Cauchy	.	.	$\frac{1}{2}$	$\frac{1}{2}$	2

logistic, exponential-gamma, Weibull-gamma) distribution [97, 108, 66]:

$$\mathrm{Burr}(x \; ; a, s, \gamma, \beta) = \frac{\beta\gamma}{|s|}\left(\frac{x-a}{s}\right)^{\beta-1}\left(1+\left(\frac{x-a}{s}\right)^{\beta}\right)^{-\gamma-1} \quad (18.3)$$

$$= \mathrm{GenBetaPrime}(x \; ; a, s, 1, \gamma, \beta)$$

Most commonly encountered as a model of income distribution.

Table 18.2: Properties of the generalized beta prime distribution

Properties		
notation	$\mathrm{GenBetaPrime}(x\,;\,a,s,\alpha,\gamma,\beta)$	
PDF	$\dfrac{1}{B(\alpha,\gamma)}\left\lvert\dfrac{\beta}{s}\right\rvert\left(\dfrac{x-a}{s}\right)^{\alpha\beta-1}\left(1+\left(\dfrac{x-a}{s}\right)^{\beta}\right)^{-\alpha-\gamma}$	
CDF / CCDF	$\dfrac{B\left(\alpha,\gamma;(1+(\frac{x-a}{s})^{-\beta})^{-1}\right)}{B(\alpha,\gamma)}$	$\frac{\beta}{s}>0\,/\,\frac{\beta}{s}<0$
	$=I\left(\alpha,\gamma;(1+(\frac{x-a}{s})^{-\beta})^{-1}\right)$	
parameters	$a,\ s,\ \alpha,\ \gamma,\ \beta$ in $\mathbb{R}$	
	$\alpha>0,\gamma>0$	
support	$x\geqslant a$	$s>0$
	$x\leqslant a$	$s<0$
mean	$a+\dfrac{sB(\alpha+\frac{1}{\beta},\gamma-\frac{1}{\beta})}{B(\alpha,\gamma)}$	$-\alpha<\frac{1}{\beta}<\gamma$
variance	$s^{2}\left[\dfrac{B(\alpha+\frac{2}{\beta},\gamma-\frac{2}{\beta})}{B(\alpha,\gamma)}-\left(\dfrac{B(\alpha+\frac{1}{\beta},\gamma-\frac{1}{\beta})}{B(\alpha,\gamma)}\right)^{2}\right]$	$-\alpha<\frac{2}{\beta}<\gamma$
skew	not simple	
ex. kurtosis	not simple	
$E[X^{h}]$	$\dfrac{\lvert s\rvert^{h}B(\alpha+\frac{h}{\beta},\gamma-\frac{h}{\beta})}{B(\alpha,\gamma)}$	$a=0,\ -\alpha<\frac{h}{\beta}<\gamma$ [51]

Dagum (Inverse Burr, Burr type III, Dagum type I, beta-kappa, beta-k, Mielke) distribution [97, 109, 108]:

$$\text{Dagum}(x \; ; \gamma, \beta) = \frac{\beta\gamma}{|s|}\left(\frac{x-a}{s}\right)^{\gamma\beta-1}\left(1+\left(\frac{x-a}{s}\right)^{\beta}\right)^{-\gamma-1} \tag{18.4}$$

$$= \text{GenBetaPrime}(x \; ; a, s, 1, \gamma, -\beta)$$

$$= \text{GenBetaPrime}(x \; ; a, s, \gamma, 1, +\beta)$$

Paralogistic distribution [66]:

$$\text{Paralogistic}(x \; ; a, s, \beta) = \frac{\beta^2}{|s|}\frac{\left(\frac{x-a}{s}\right)^{\beta-1}}{(1+\left(\frac{x-a}{s}\right)^{\beta})^{\beta+1}} \tag{18.5}$$

$$= \text{GenBetaPrime}(x \; ; a, s, 1, \beta, \beta)$$

Inverse paralogistic distribution [107]:

$$\text{InvParalogistic}(x \; ; a, s, \beta) = \frac{\beta^2}{|s|}\frac{\left(\frac{x-a}{s}\right)^{\beta^2-1}}{(1+\left(\frac{x-a}{s}\right)^{\beta})^{\beta+1}} \tag{18.6}$$

$$= \text{GenBetaPrime}(x \; ; a, s, \beta, 1, \beta)$$

Log-logistic (Fisk, Weibull-exponential, Pareto type III, power prime) distribution [110, 3]:

$$\text{LogLogistic}(x \; ; a, s, \beta) = \left|\frac{\beta}{s}\right|\frac{\left(\frac{x-a}{s}\right)^{\beta-1}}{\left(1+\left(\frac{x-a}{s}\right)^{\beta}\right)^2} \tag{18.7}$$

$$= \text{Burr}(x \; ; a, s, 1, \beta)$$

$$= \text{GenBetaPrime}(x \; ; 0, s, 1, 1, \beta)$$

Used as a parametric model for survival analysis and, in economics, as a model for the distribution of wealth or income. The logistic and log-logistic distributions are related by an exponential transform.

$$\text{LogLogistic}(0, s, \beta) \sim \exp\left(-\text{Logistic}(-\ln s, \tfrac{1}{\beta})\right)$$

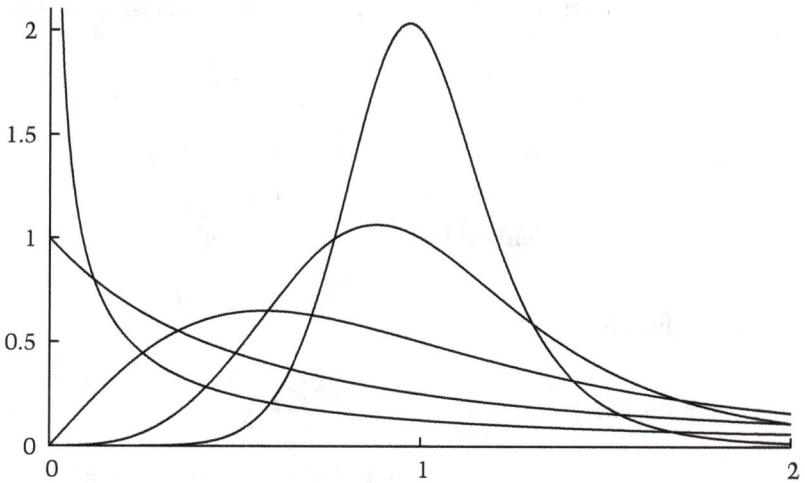

Figure 37: Log-logistic distributions, $\mathrm{LogLogistic}(x\,;\,0,1,\beta)$.

Half-Pearson VII (half-t) distribution [111]:

$$\mathrm{HalfPearsonVII}(x\,;\,a,s,m) \tag{18.8}$$

$$= \frac{1}{B(\frac{1}{2}, m - \frac{1}{2})} \frac{2}{|s|} \left(1 + \left(\frac{x-a}{s}\right)^2\right)^{-m}$$

$$= \mathrm{GenBetaPrime}(x\,;\,a,s,\tfrac{1}{2}, m - \tfrac{1}{2}, 2)$$

The Pearson type VII (9.1) distribution truncated at the center of symmetry. Investigated as a prior for variance parameters in hierarchal models [111].

Half-Cauchy distribution [111]:

$$\mathrm{HalfCauchy}(x\,;\,a,s) = \frac{2}{\pi|s|} \left(1 + \left(\frac{x-a}{s}\right)^2\right)^{-1} \tag{18.9}$$

$$= \mathrm{HalfPearsonVII}(x\,;\,a,s,1)$$

$$= \mathrm{GenBetaPrime}(x\,;\,a,s,\tfrac{1}{2}, \tfrac{1}{2}, 2)$$

A notable subclass of the Half-Pearson type VII, the Cauchy distribution

(9.6) truncated at the center of symmetry.

Half generalized Pearson VII distribution [1]:

$$\text{HalfGenPearsonVII}(x \ ; \ a, s, m, \beta) \qquad\qquad (18.10)$$

$$= \frac{\beta}{|s|B(m - \frac{1}{\beta}, \frac{1}{\beta})} \left(1 + \left(\frac{x-a}{s}\right)^{\beta}\right)^{-m}$$

$$= \text{GenBetaPrime}(x \ ; \ a, s, \tfrac{1}{\beta}, m - \tfrac{1}{\beta}, \beta)$$

One half of a Generalized Pearson VII distribution (21.6). Special cases include half Pearson VII (18.8), half Cauchy (18.9), **half-Laha** (See (20.18)), and uniform prime (5.8) distributions.

$$\text{HalfGenPearsonVII}(x \ ; \ a, s, m, 2) = \text{HalfPearsonVII}(x \ ; \ a, s, m)$$

$$\text{HalfGenPearsonVII}(x \ ; \ a, s, 1, 2) = \text{HalfCauchy}(x \ ; \ a, s)$$

$$\text{HalfGenPearsonVII}(x \ ; \ a, s, 1, 4) = \text{HalfLaha}(x \ ; \ a, s)$$

$$\text{HalfGenPearsonVII}(x \ ; \ a, s, 2, 1) = \text{UniPrime}(x \ ; \ a, s)$$

The half exponential power (11.4) distribution occurs in the large m limit.

$$\lim_{m \to \infty} \text{HalfGenPearsonVII}(x \ ; \ a, \theta m^{\frac{1}{\beta}}, m, \beta) = \text{HalfExpPower}(x \ ; \ a, \theta, \beta)$$

Interrelations

Negating the Weibull parameter of the generalized beta prime distribution is equivalent to exchanging the shape parameters α and γ.

$$\text{GenBetaPrime}(x \ ; \ a, s, \alpha, \gamma, \beta) = \text{GenBetaPrime}(x \ ; \ a, s, \gamma, \alpha, -\beta)$$

The distribution is related to ratios of gamma distributions.

$$\text{GenBetaPrime}(a, s, \alpha, \gamma, \beta) \sim a + s \left(\frac{\text{StdGamma}_1(\alpha)}{\text{StdGamma}_2(\gamma)}\right)^{\frac{1}{\beta}}$$

Limit of the generalized beta prime distribution include the Amoroso

(11.1) [51] and beta-logistic (15.1) distributions.

$$\lim_{\gamma \to \infty} \text{GenBetaPrime}(x \; ; \; a, \theta\gamma^{\frac{1}{\beta}}, \alpha, \gamma, \beta) = \text{Amoroso}(x \; ; \; a, \theta, \alpha, \beta)$$

$$\lim_{\beta \to \infty} \text{GenBetaPrime}(x \; ; \; \zeta + \beta\lambda, -\beta\lambda, \alpha, \gamma, \beta) = \text{BetaLogistic}(x \; ; \; \zeta, \lambda, \gamma, \alpha)$$

Therefore, the generalized beta prime also indirectly limits to the normal (4.1), log-normal (6.1), gamma-exponential (8.1), Laplace (3.1) and power-function (5.1) distributions, among others.

Generalized beta prime describes the order statistics (§C) of the log-logistic distribution (18.7)).

$$\text{OrderStatistic}_{\text{LogLogistic}(a,s,\beta)}(x \; ; \; \gamma, \alpha) = \text{GenBetaPrime}(x \; ; \; a, s, \alpha, \gamma, \beta)$$

Despite occasional claims to the contrary, the log-Cauchy distribution is not a special case of the generalized beta prime distribution (generalized beta prime is mono-modal, log-Cauchy is not).

19   Pearson Distribution

The **Pearson** distributions [5, 6, 7, 112, 2] are a family of continuous, univariate, unimodal probability densities with distribution function

$$\text{Pearson}(x \;;\; a, s, \;\; a_1, a_2, \;\; b_0, b_1, b_2) \tag{19.1}$$

$$= \tfrac{1}{\mathcal{N}} \left(1 - \tfrac{1}{r_0}\tfrac{x-a}{s}\right)^{e_0} \left(1 - \tfrac{1}{r_1}\tfrac{x-a}{s}\right)^{e_1}$$

$$a, \; s, \; a_1, \; a_2, \; b_0, \; b_1, \; b_2, \; x \; \text{in} \; \mathbb{R}$$

$$r_0 = \frac{-b_1 + \sqrt{b_1^2 - 4b_2 b_0}}{2b_2} \qquad e_0 = \frac{-a_1 - a_2 r_0}{r_1 - r_0}$$

$$r_1 = \frac{-b_1 - \sqrt{b_1^2 - 4b_2 b_0}}{2b_2} \qquad e_1 = \frac{a_1 + a_2 r_1}{r_1 - r_0}$$

Here $\mathcal{N}$ is the normalization constant. Note that the parameter a_2 is redundant, and can be absorbed into the scale. Thus the Pearson distribution effectively has 4 shape parameters. We retain a_2 in the general definition since this makes parameterization of subtypes easier.

Pearson constructed his family of distributions by requiring that they satisfy the differential equation

$$\frac{d}{dx} \ln \text{Pearson}(x \;;\; 0, 1, \;\; a_1, a_2, \;\; b_0, b_1, b_2) = -\frac{a_1 + a_2 x}{b_0 + b_1 x + b_2 x^2} \;,$$

$$= -\frac{1}{x}\frac{a_1 x + a_2 x^2}{b_0 + b_1 x + b_2 x^2} \;,$$

$$= \frac{e_0}{x - r_0} + \frac{e_1}{x - r_1} \;.$$

Pearson's original motivation was that the discrete hypergeometric distribution obeys an analogous finite difference relation [112], and that at the time very few continuous, univariate, unimodal probability distributions had been described. The numbering of the a_1, a_2 coefficients is chosen to be consistent with Weibull transformed generalization of the Pearson distribution (20.1), where an a_0 parameter naturally arises.

The Pearson distribution has three main subtypes determined by r_0 and r_1, the roots of the quadratic denominator. First, we can have two roots located on the real line, at the minimum and maximum of the distribution. This is commonly known as the beta distribution (12.1). (The parameterization is based on standard conventions.)

$$p(x) \propto x^{\alpha-1}(1-x)^{\gamma-1}, \qquad 0 < x < 1$$

The second possibility is that the distribution has semi infinite support, with one root at the boundary, and the other located outside the distribution's support. This is the beta prime distribution. (13.1) (Again, the parameterization is based on standard conventions.)

$$p(x) \propto x^{\alpha-1}(1+x)^{-\alpha-\gamma}, \qquad 0 < x < +\infty$$

The third possibility is that the distribution has an infinite support with both roots located off the real axis in the complex plane. To ensure that the distribution remains real, the roots must be complex conjugates of one another. In this case, the root order can also be complex conjugates of one another. This is Pearson's type IV distribution (16.1). (The complex roots and powers can be disguised with trigonometric functions and some algebra, at the cost of making the distribution look more complex than it actually is.)

$$p(x) \propto (i-x)^{m+iv}(i+x)^{m-iv}, \qquad -\infty < x < +\infty$$

The Cauchy distribution, for instance, is a special case of Pearson's type IV distribution.

Special cases

A large number of useful distributions are members of Pearson's family (See Fig. 2). Pearson identified 13 principal subtypes – the normal distribution and types I through XII (See table 19.1). In Fig. 2 and table 19.2 we consider 12 principal subtypes. (We include the uniform, inverse exponential and Cauchy as distributions important in their own right, and give less prominence to Pearson's types VIII, IX, XI and XII.) All of the Pearson distributions have great utility and are widely applied, with the exception of Pearson IV (infinite support, complex roots with complex powers) (16.1), which appears rarely (if at all) in practical applications.

q-Gaussian (symmetric Pearson) distribution [113] :

$$\text{QGaussian}(x \; ; \mu, \sigma, q) = \frac{1}{\sqrt{2\sigma^2}\,\mathcal{N}} \exp_q\left(-\tfrac{1}{2}\left(\tfrac{x-\mu}{\sigma}\right)^2\right) \qquad (19.2)$$

$$= \frac{1}{\sqrt{2\sigma^2}\,\mathcal{N}}\left(1 - \tfrac{1}{2}(1-q)\left(\tfrac{x-\mu}{\sigma}\right)^2\right)^{\frac{1}{1-q}}$$

$$-2 < q < 3$$

$$x \in (-\infty, +\infty) \text{ for } 1 \leqslant q < 3$$

$$x \in \left(\mu - \tfrac{\sqrt{2}\sigma}{\sqrt{1-q}}, \mu + \tfrac{\sqrt{2}\sigma}{\sqrt{1-q}}\right) \text{ for } q < 1$$

Here $\exp_q$ is the q-generalized exponential function (§F). The normalization constant is

$$\mathcal{N} = \begin{cases} \sqrt{\pi}\dfrac{2\Gamma\left(\frac{1}{1-q}\right)}{(3-q)\sqrt{1-q}\,\Gamma\left(\frac{3-q}{2(1-q)}\right)} & -2 < q < +1 \\[2ex] \sqrt{\pi} & q = +1 \\[2ex] \sqrt{\pi}\dfrac{\Gamma\left(\frac{3-q}{2(q-1)}\right)}{\sqrt{q-1}\,\Gamma\left(\frac{1}{q-1}\right)} & +1 < q < +3 \end{cases}$$

A special case of the Pearson family that interpolates between all of the symmetric Pearson distributions: the central beta (12.5), normal (4.1) and Pearson VII (9.1) families. See also the hierarchy of symmetric distributions in Fig. 5.

$$\text{QGaussian}(x \; ; \mu, \sigma, q)$$

$$= \begin{cases} \text{Beta}\left(x \; ; a - \tfrac{\sqrt{2}\sigma}{\sqrt{1-q}}, \tfrac{2\sqrt{2}\sigma}{\sqrt{1-q}}, \tfrac{2-q}{1-q}, \tfrac{2-q}{1-q}\right) & -2 < q < 1 \\[1.5ex] \text{CentralBeta}\left(x \; ; a, \tfrac{\sqrt{2}\sigma}{\sqrt{1-q}}, \tfrac{2-q}{1-q}\right) & -2 < q < 1 \\[1.5ex] \text{Normal}(x \; ; \mu, \sigma) & q = 1 \\[1.5ex] \text{PearsonVII}\left(x \; ; a, \tfrac{\sqrt{2}\sigma}{\sqrt{q-1}}, \tfrac{1}{q-1}\right) & 1 < q < 3 \end{cases}$$

Table 19.1: Pearson's categorization

type	notes	Eq.	Ref.
	normal	(4.1)	[5]
I	beta	(12.1)	[5]
II	central-beta	(12.5)	[5]
III	gamma	(7.1)	[4]
IV	Includes Pearson VII	(16.1)	[5]
V	inverse gamma	(11.13)	[6]
VI	beta prime	(13.1)	[6]
VII	Includes Cauchy and Student's t	(9.1)	[7]
VIII	Special case of power function	(5.1)	[7]
IX	Special case of power function	(5.1)	[7]
X	exponential	(2.1)	[7]
XI	Pareto	(5.5)	[7]
XII	J-shaped beta	(12.4)	[7]

Table 19.2: Special cases of the Pearson distribution

(19.1)	Pearson	a	s	a_1	a_2	b_0	b_1	b_2
(1.1)	uniform	a	s	0	0	0	1	-1
(12.5)	central-beta	$\mu\text{-}b$	$2b$	$\alpha - 1$	$2\alpha - 2$	0	1	-1
(12.1)	beta	a	s	$\alpha - 1$	$\alpha + \gamma - 2$	0	1	-1
(2.1)	exponential	a	θ	0	-1	0	1	0
(7.1)	gamma	a	θ	$\alpha - 1$	-1	0	1	0
(13.1)	beta-prime	a	s	$\alpha - 1$	$-\gamma - 1$	0	1	1
(11.13)	inv. gamma	a	θ	-1	$\alpha + 1$	0	0	1
(11.14)	inv. exp.	a	θ	-1	2	0	0	1
(16.1)	Pearson IV	a	s	2ν	$2m$	1	0	1
(9.1)	Pearson VII	a	s	0	$2m$	1	0	1
(9.6)	Cauchy	a	s	0	2	1	0	1
(4.1)	normal	μ	σ	0	2	1	0	0

20 GRAND UNIFIED DISTRIBUTION

The Grand Unified Distribution of order n is required to satisfy the following differential equation.

$$\frac{d}{dx} \ln \mathrm{GUD}^{(n)}(x \; ; \; a, s, \; a_0, a_1, \ldots, a_n, \; b_0, b_1, \ldots, b_n, \; \beta) \qquad (20.1)$$

$$= -\left|\frac{\beta}{s}\right| \frac{1}{\left(\frac{x-a}{s}\right)} \frac{a_0 + a_1\left(\frac{x-a}{s}\right)^{\beta} + \cdots + a_n\left(\frac{x-a}{s}\right)^{n\beta}}{b_0 + b_1\left(\frac{x-a}{s}\right)^{\beta} + \cdots + b_n\left(\frac{x-a}{s}\right)^{n\beta}}$$

$$a, \; s, \; a_0, a_1, \ldots, a_n, \; b_0, b_1, \ldots, b_n, \; \beta, \; x \; \text{in} \; \mathbb{R}$$

$$\beta = 1 \; \text{when} \; a_0 = 0$$

In principal, any analytic probability distribution can satisfy this relation. The central hypothesis of this compendium is that most interesting univariate continuous probability distributions satisfy this relation with low order polynomials in the denominator and numeration. If fact, there seems be little need to consider beyond $n = 2$, which we take as the default order, in the absence of further qualification.

Special cases

Extended Pearson distribution [114]: With $\beta = 1$ we obtain an extended Pearson distribution.

$$\frac{d}{dx} \ln \mathrm{ExtPearson}(x \; ; 0, 1, \; a_0, a_1, a_2, \; b_0, b_1, b_2) \qquad (20.2)$$

$$= -\frac{1}{x} \frac{a_0 + a_1 x + a_2 x^2}{b_0 + b_1 x + b_2 x^2}$$

$$a, \; s, \; a_0, \; a_1, \; a_2, \; b_0, \; b_1, \; b_2 \; \text{in} \; \mathbb{R}$$

Inverse Gaussian (Wald, inverse normal) distribution [115, 116, 117, 118, 2]:

$$\mathrm{InvGaussian}(x \; ; \; \mu, \lambda) = \sqrt{\frac{\lambda}{2\pi x^3}} \exp\left(\frac{-\lambda(x - \mu)^2}{2\mu^2 x}\right) \qquad (20.3)$$

$$= \mathrm{ExtPearson}(x \; ; 0, 1, \; -\tfrac{\lambda}{2}, \tfrac{3}{2}, \tfrac{\lambda}{2\mu^2}, \; 0, 1, 0)$$

$$= \mathrm{GUD}(x \; ; 0, 1, \; -\tfrac{\lambda}{2}, \tfrac{3}{2}, \tfrac{\lambda}{2\mu^2}, \; 0, 1, 0, \; 1)$$

Figure 38: Grand Unified Distributions

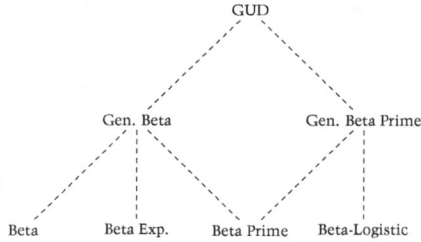

Table 20.1: Special cases of the Grand Unified Distribution

	GUD	a	s	a_0	a_1	a_2	b_0	b_1	b_2	β
(20.1)	GUD	.	.	.	.	.	.	.	.	
(20.2)	Ext. Pearson	.	.	.	.	.	.	.	.	1
(19.1)	Pearson	.	.	0	.	.	.	.	.	1
(17.1)	gen. beta	.	.	.	.	0	0	1	-1	.
(17.1)	gen. beta prime	.	.	.	.	0	0	1	1	.
(20.3)	inv. Gaussian	.	.	.	$\frac{3}{2}$	.	0	1	0	1
(20.4)	rec. inv. Gaussian	.	.	.	$\frac{3}{2}$	.	0	1	0	-1
(20.5)	Halphen	.	.	$-\kappa$	$1-\alpha$	κ	0	1	0	1
(20.13)	gen. Halphen	.	.	$-\kappa$	$1-\alpha$	κ	0	1	0	β
(20.6)	Hyperbola	.	.	$-\kappa$	1	κ	0	1	0	1
(20.7)	Halphen B	.	.	$1-\alpha$	$-\kappa$	2	1	0	0	1
(20.8)	inv. Halphen B	.	.	-2	$-\kappa$	$1-\alpha$	0	0	1	1
(20.9)	Sichel	.	.	$-\lambda$	$1-\alpha$	κ	0	1	0	1
(20.14)	gen. Sichel	.	.	$-\lambda$	$1-\alpha$	κ	0	1	0	β

with support $x > 0$, mean $\mu > 0$, and shape $\lambda > 0$. The name 'inverse Gaussian' is misleading, since this is not in any direct sense the inverse of a Gaussian distribution. The **Wald** distribution is a special case with $\mu = 1$.

The inverse Gaussian distribution describes first passage time in one dimensional Brownian diffusion with drift [118]. The displacement x of a diffusing particle after a time t, with diffusion constant D and drift velocity v, is Normal$(vt, \sqrt{2Dt})$. The 'inverse' problem is to ask for the first passage time, the time taken to first reach a particular position $y > 0$, which is distributed as InvGaussian$(\frac{y}{v}, \frac{y^2}{2D})$.

In the limit that μ goes to infinity we recover the Lévy distribution (11.15), the first passage time distribution for Brownian diffusion without drift.

$$\lim_{\mu \to \infty} \text{InvGaussian}(x \; ; \mu, \lambda) = \text{Lévy}(x \; ; 0, \lambda)$$

The sum of independent inverse Gaussian random variables is also inverse Gaussian, provided that μ^2/λ is a constant.

$$\sum_i \text{InvGaussian}_i(x \; ; \mu' w_i, \lambda' w_i^2)$$

$$\sim \text{InvGaussian}\left(x \; ; \mu' \sum_i w_i, \lambda'\left(\sum_i w_i\right)^2\right)$$

Scaling an inverse Gaussian scales both μ and λ.

$$c \; \text{InvGaussian}(\mu, \lambda) \sim \text{InvGaussian}(c\mu, c\lambda)$$

It follows from the previous two relations the sample mean of an inverse Gaussian is inverse Gaussian.

$$\frac{1}{N} \sum_{i=1}^{N} \text{InvGaussian}_i(\mu, \lambda) \sim \text{InvGaussian}(\mu, N\lambda)$$

Reciprocal inverse Gaussian distribution [2]:

$$\text{RecInvGaussian}(x \; ; \mu, \lambda) = \sqrt{\frac{\lambda}{2\pi x}} \exp\left(\frac{-\lambda(1 - \mu x)^2}{2\mu^2 x}\right) \qquad (20.4)$$

$$= \text{ExtPearson}(x \; ; 0, 1, \; -\tfrac{\lambda}{2\mu^2}, \tfrac{1}{2}, \tfrac{\lambda}{2}, \; 0, 1, 0)$$

$$= \text{GUD}(x \; ; 0, 1, \; -\tfrac{\lambda}{2}, \tfrac{3}{2}, \tfrac{\lambda}{2\mu^2}, \; 0, 1, 0, \; -1)$$

with support $x > 0$, mean $\mu > 0$, and shape $\lambda > 0$. An inverted (in standard sense) inverse Gaussian distribution.

$$\text{RecInvGaussian}(\mu, \lambda) \sim \text{InvGaussian}(\mu, \lambda)^{-1}$$

Halphen (Halphen A) distribution [119]:

$$\text{Halphen}(x \; ; a, s, \alpha, \kappa) \qquad (20.5)$$

$$= \frac{1}{2|s|K_\alpha(2\kappa)} \left(\frac{x - a}{s}\right)^{\alpha - 1} \exp\left\{-\kappa\left(\frac{x - a}{s}\right) - \kappa\left(\frac{x - a}{s}\right)^{-1}\right\},$$

$$= \text{GUD}(x \; ; a, s, \; -\kappa, 1 - \alpha, \kappa, \; 0, 1, 0, \; 1)$$

$$0 \leqslant \tfrac{x-a}{s}$$

Developed by Étienne Halphen for the frequency analysis of river flows. Limits to gamma, inverse gamma, and normal.

Hyperbola (harmonic) distribution [119, 120]:

$$\text{Hyperbola}(x \; ; a, s, \kappa) \qquad (20.6)$$

$$= \frac{1}{2|s|K_0(2\kappa)} \left(\frac{x - a}{s}\right)^{-1} \exp\left\{-\kappa\left(\frac{x - a}{s}\right) - \kappa\left(\frac{x - a}{s}\right)^{-1}\right\},$$

$$= \text{Halphen}(x \; ; a, s, 0, \kappa)$$

$$= \text{GUD}(x \; ; a, s, \; -\kappa, 1, \kappa, \; 0, 1, 0, \; 1)$$

$$0 \leqslant \tfrac{x-a}{s}$$

Halphen B distribution [119, 120]:

$$\text{HalphenB}(x \; ; a, s, \alpha, \kappa) \tag{20.7}$$
$$= \frac{2}{|s| H_{2\alpha}(\kappa)} \left(\frac{x-a}{s}\right)^{\alpha-1} \exp\left\{-\left(\frac{x-a}{s}\right)^2 + \kappa\left(\frac{x-a}{s}\right)\right\},$$
$$= \text{GUD}(x \; ; a, s, \; 1-\alpha, -\kappa, 2, \; 1, 0, 0, \; 1)$$
$$0 \leqslant \tfrac{x-a}{s}$$

The normalizing function $H_{2\alpha}(\kappa)$ was called the exponential factorial function by Halphen [121, 120]. Limits to gamma distribution (7.1) as $\kappa \to \infty$.

Inverse Halphen B distribution [122, 120]:

$$\text{InvHalphenB}(x \; ; a, s, \alpha, \kappa) \tag{20.8}$$
$$= \frac{2}{|s| H_{2\alpha}(\kappa)} \left(\frac{x-a}{s}\right)^{-\alpha+1} \exp\left\{-\left(\frac{x-a}{s}\right)^{-2} + \kappa\left(\frac{x-a}{s}\right)^{-1}\right\},$$
$$= \text{GUD}(x \; ; a, s, \; -2, -\kappa, 1-\alpha, \; 0, 0, 1, \; 1)$$
$$0 \leqslant \tfrac{x-a}{s}$$

Limits to inverse gamma distribution (11.13) as $\kappa \to \infty$.

Sichel (generalized inverse Gaussian) distribution [123, 124, 125]:

$$\text{Sichel}(x \; ; a, s, \alpha, \kappa, \lambda) \tag{20.9}$$
$$= \frac{(\kappa/\lambda)^{\alpha/2}}{2|s| K_\alpha(2\sqrt{\kappa\lambda})} \left(\frac{x-a}{s}\right)^{\alpha-1} \exp\left\{-\kappa\left(\frac{x-a}{s}\right) - \lambda\left(\frac{x-a}{s}\right)^{-1}\right\},$$
$$= \text{GUD}(x \; ; a, s, \; -\lambda, 1-\alpha, \kappa, \; 0, 1, 0, \; 1)$$
$$0 \leqslant \tfrac{x-a}{s}$$

Special cases include Halphen (20.5) $\lambda = \kappa$, and inverse Gaussian (20.3) $\alpha = -\tfrac{1}{2}$.

Libby-Novick distribution [126, 127, 128, 129]

$$\text{LibbyNovick}(x \; ; \; a, s, c, \alpha, \gamma) \qquad\qquad (20.10)$$
$$= \frac{1}{|s|B(\alpha, \gamma)} \left(\tfrac{x-a}{s}\right)^{\alpha-1} \left(1 - \tfrac{x-a}{s}\right)^{\gamma-1} \left(1 - (1-c)\tfrac{x-a}{s}\right)^{-\alpha-\gamma}$$
$$= \text{GUD}(x \; ; \; a, s, \; \alpha-1, 3-\alpha-c-c\gamma, 2c-2,$$
$$1, c-2, 1-c, \; 1)$$
$$\text{for } a, s, c, \alpha, \gamma \text{ in } R, \alpha, \gamma > 0$$
$$0 \leqslant \tfrac{x-a}{s} \leqslant 1$$

A generalized three-parameter beta distribution that arises naturally as a beta distribution style ratio of gamma distributions [128].

$$\text{LibbyNovick}(0, \tfrac{s_1}{s_2}, \alpha, \gamma) \sim \frac{\text{Gamma}_1(0, s_1, \alpha)}{\text{Gamma}_1(0, s_1, \alpha) + \text{Gamma}_2(0, s_2, \gamma)}$$

Limits to both the beta ($u = 1$) and beta-prime ($u \to \infty$) distributions.

Gauss hypergeometric distribution [130, 128]

$$\text{GaussHypergeometric}(x \; ; \; a, s, u, \alpha, \gamma, \delta) \qquad\qquad (20.11)$$
$$= \frac{1}{|s|\mathcal{N}} \left(\frac{x-a}{s}\right)^{\alpha-1} \left(1 - \frac{x-a}{s}\right)^{\gamma-1} \left(1 - (1-u)\frac{x-a}{s}\right)^{-\delta}$$
$$\mathcal{N} = B(\alpha, \gamma) \, {}_2F_1(\alpha, \delta; \alpha+\gamma, 1-u)$$
$$\text{for } a, s, u, \alpha, \gamma, \delta \text{ in } \mathbb{R}, \alpha, \gamma, \delta > 0$$
$$= \text{GUD}(x \; ; \; a, s, \; \alpha-1, 2-\alpha-\gamma+(1-u)(1+\rho+\alpha),$$
$$u(\alpha+\gamma-\rho-2), \; 1, -1-c, -u, \; 1)$$
$$0 \leqslant \tfrac{x-a}{s} \leqslant 1$$

Motivated by the Euler integral formula for the Gauss hypergeometric function (§F).

Confluent hypergeometric distribution [131, 132, 129]

$$\text{Confluent}(x\ ;\ \alpha,\gamma,\delta) \tag{20.12}$$

$$= \frac{1}{\mathcal{N}}\left(\frac{x-a}{s}\right)^{\alpha-1}\left(1-\left(\frac{x-a}{s}\right)\right)^{\gamma-1}\exp\left\{-\kappa\left(\frac{x-a}{s}\right)\right\}$$

$$\mathcal{N} = B(\alpha,\gamma)\ {}_1F_1(\alpha;\alpha+\gamma;-\kappa)$$

$$= \text{GUD}(x\ ;\ 0,1,\ 1-\alpha,\alpha+\gamma+\kappa-2,-\kappa,\ 1,-1,0,\ 1)$$

$$0 \leqslant \tfrac{x-a}{s} \leqslant 1$$

This distribution was introduced by Gordy [131] for applications to auction theory.

Generalized Halphen [1] :

$$\text{GenHalphen}(x\ ;\ a,s,\alpha,\kappa,\beta) \tag{20.13}$$

$$= \frac{|\beta|}{2|s|K_\alpha(2\kappa)}\left(\frac{x-a}{s}\right)^{\beta\alpha-1}\exp\left\{-\kappa\left(\frac{x-a}{s}\right)^\beta - \kappa\left(\frac{x-a}{s}\right)^{-\beta}\right\}$$

$$= \text{GUD}(x\ ;\ a,s,\ -\kappa,1-\alpha,\kappa,\ 0,1,0,\ \beta)$$

$$0 \leqslant \left(\tfrac{x-a}{s}\right)^\beta$$

Generalized Sichel (generalized generalized inverse Gaussian) distribution [70]:

$$\text{GenSichel}(x\ ;\ a,s,\alpha,\kappa,\lambda,\beta) \tag{20.14}$$

$$= \frac{|\beta|(\kappa/\lambda)^{\alpha/2}}{2|s|K_\alpha(2\sqrt{\kappa\lambda})}\left(\frac{x-a}{s}\right)^{\beta\alpha-1}\exp\left\{-\kappa\left(\frac{x-a}{s}\right)^\beta - \lambda\left(\frac{x-a}{s}\right)^{-\beta}\right\},$$

$$= \text{GUD}(x\ ;\ a,s,\ -\lambda,1-\alpha,\kappa,\ 0,1,0,\ \beta)$$

$$0 \leqslant \left(\tfrac{x-a}{s}\right)^\beta$$

Special cases include the generalized Halphen (20.13) $\lambda = \kappa$, and Sichel (20.9) distributions $\beta = 1$.

Table 20.2: Special cases of the Pearson exponential family

(20.15)	Pearson Exp.	ζ	λ	a_0	a_1	a_2	b_0	b_1	b_2
(14.1)	beta-exp.	.	.	0	$\alpha+\gamma$-1	$-\alpha$	0	1	-1
(15.1)	beta-logistic	.	.	0	$-\gamma$	α	0	1	1
(15.4)	central-logistic	.	.	0	$-\alpha$	α	0	1	1
(20.16)	Perks	.	.	-1	0	1	1	c	1
(15.5)	logistic	.	.	0	-1	1	0	1	1
(15.6)	hyperbolic secant	.	.	0	$-\frac{1}{2}$	$\frac{1}{2}$	0	1	1
(8.1)	gamma exp.	.	.	0	$-\alpha$	1	0	1	0
(2.1)	exponential	.	.	0	1	0	0	1	0

Pearson-exponential distributions

If we take the limit of β to infinity (See (§D)), then we get the family of Pearson exponential distributions.

Pearson-exponential distribution [1]:

$$\text{PearsonExp}(x \; ; \zeta, \lambda, \quad a_0, a_1, a_2, \quad b_0, b_1, b_2)$$
$$= \lim_{\beta \to \infty} \text{GUD}(x \; ; \zeta + \beta\lambda, \beta\lambda, \quad a_0, a_1, a_2, \quad b_0, b_1, b_2, \quad \beta)$$

Because we can generally interchange limits and differentiation, such distributions satisfy the following differential equation.

$$\frac{d}{dx} \ln \text{PearsonExp}(x \; ; \zeta, \lambda, \quad a_0, a_1, a_2, \quad b_0, b_1, b_2)$$

$$= \left| \frac{1}{\lambda} \right| \frac{a_0 + a_1 e^{\frac{x-\zeta}{\lambda}} + a_2 e^{2\frac{x-\zeta}{\lambda}}}{b_0 + b_1 e^{\frac{x-\zeta}{\lambda}} + b_2 e^{2\frac{x-\zeta}{\lambda}}}$$

See table 20.2 and Fig. 3.

Perks (Champernowne) distribution [100, 133, 101, 42]:

$$\text{Perks}(x \; ; \zeta, \lambda, c) = \frac{1}{\mathcal{N}} \frac{1}{c + e^{-\frac{x-\zeta}{\lambda}} + e^{+\frac{x-\zeta}{\lambda}}}$$

$$= \text{PearsonExp}(x \; ; \zeta, \lambda, \quad -1, 0, -1, \quad 1, c, 1)$$

Special cases include logistic ($c = 0$) (15.5) and hyperbolic secant ($c = 2$) (15.6) distributions.

Greater Grand Unified distributions

There are only a few interesting specials cases of the Grand Unified Distribution with order greater than 2.

Appell Beta distribution [132]:

$$
\begin{aligned}
&\text{AppellBeta}(x \; ; \; a, s, \alpha, \gamma, \rho, \delta) \\
&= \frac{1}{\mathcal{N} \, |s|} \frac{\left(\frac{x-a}{s}\right)^{\alpha-1}\left(1 - \frac{x-a}{s}\right)^{\gamma-1}}{\left(1 - u\frac{x-a}{s}\right)^{\rho}\left(1 - v\frac{x-a}{s}\right)^{\delta}} \\
&\mathcal{N} = B(\alpha, \gamma) \, F_1(\alpha, \rho, \delta, \alpha + \gamma; u, v) \\
&= \text{GUD}^{(3)}(x \; ; \; a, s, \quad a_0, a_1, a_2, a_3, \quad b_0, b_1, b_2, b_3, \quad 1) \\
&b_0 = -1, \; b_1 = 1 + u + v, \; b_2 = -u - v - uv, \; b_3 = uv
\end{aligned}
\tag{20.17}
$$

Here F_1 is the Appell hypergeometric function of the first kind.

Laha distribution [134, 135, 136]:

$$
\begin{aligned}
\text{Laha}(x \; ; \; a, s) &= \frac{\sqrt{2}}{|s| \, \pi} \frac{1}{\left(1 + \left(\frac{x-a}{s}\right)^4\right)} \\
&= \text{GUD}^{(4)}(x \; ; \; a, s, \quad 0, -4, 0, 0, 0, \quad 1, 0, 2, 0, 1, \quad 1)
\end{aligned}
\tag{20.18}
$$

A symmetric, continuous, univariate, unimodal probability density, with infinite support. Originally introduced to disprove the belief that the ratio of two independent and identically distributed random variables is distributed as Cauchy (9.6) if, and only if, the distribution is normal. A 4th order Grand Unified Distribution (§20), and a special case of the generalized Pearson VII distribution (21.6).

In contradiction to the literature [136], Laha random variates can be easily generated by noting that the distribution is symmetric, and that the half-Laha distribution (18.10) is a special case of the generalized beta prime distribution, which can itself be generated as the ratio of two gamma distributions [1].

21   Miscellaneous Distributions

In this section we detail various related distributions that do not fall into the previously discussed families; either because they are not continuous, not univariate, not unimodal, or simply not simple. The notation is less uniform in this section and we do not provide detailed properties for each distribution, but instead list a few pertinent citations.

Bates distribution [137, 3]:

$$\text{Bates}(n) \sim \frac{1}{n} \sum_{i=1}^{n} \text{Uniform}_i(0,1) \qquad (21.1)$$

$$\sim \frac{1}{n} \text{IrwinHall}(n)$$

The mean of n independent standard uniform variates.

Beta-Fisher-Tippett (generalized beta-exponential) distribution [1]:

$$\text{BetaFisherTippett}(x \; ; \; \zeta, \lambda, \alpha, \gamma, \beta) \qquad (21.2)$$

$$= \frac{1}{B(\alpha,\gamma)} \left| \frac{\beta}{\lambda} \right| \left(\frac{x-\zeta}{\lambda} \right)^{\beta-1} e^{-\alpha(\frac{x-\zeta}{\lambda})^{\beta}} \left(1 - e^{-(\frac{x-\zeta}{\lambda})^{\beta}} \right)^{\gamma-1}$$

for x, ζ, λ, α, γ, β in $\mathbb{R}$,

$\alpha, \gamma > 0$, $\frac{x-\zeta}{\lambda} > 0$

A five parameter, continuous, univariate probability density, with semi-infinite support. The Beta-Fisher-Tippett occurs as the weibullization of the beta-exponential distribution (14.1), and as the order statistics of the Fisher-Tippett distribution (11.25).

$$\text{OrderStatistic}_{\text{FisherTippett}(a,s,\beta)}(x \; ; \; \alpha, \gamma)$$

$$= \text{BetaFisherTippett}(x \; ; \; a, s, \alpha, \gamma, \beta)$$

The order statistics of the Weibull (11.27) and Fréchet (11.29) distributions are therefore also Beta-Fisher-Tippett.

With $\beta = 1$ we recover the beta-exponential distribution (14.1). Other special cases include the **inverse beta-exponential**, $\beta = -1$ [1] (The order statistics of the inverse exponential distribution, (11.14)), and the **exponentiated Weibull** distribution, $\alpha = 1$ [138, 139].

Birnbaum-Saunders (fatigue life distribution) distribution [140, 3]:

$$\mathrm{BirnbaumSaunders}(x \; ; \; a, s, \gamma) \tag{21.3}$$

$$= \frac{1}{2\gamma\sqrt{2\pi s^2}} \frac{s}{x-a} \left(\sqrt{\frac{x-a}{s}} + \sqrt{\frac{s}{x-a}} \right) \exp\left\{ \frac{\left(\sqrt{\frac{x-a}{s}} - \sqrt{\frac{s}{x-a}} \right)^2}{2\gamma^2} \right\}$$

Models physical fatigue failure due to crack growth.

Exponential power (Box-Tiao, generalized normal, generalized error, Subbotin) distribution [141, 142]:

$$\mathrm{ExpPower}(x \; ; \; \zeta, \theta, \beta) = \frac{\beta}{2|\theta|\Gamma(\frac{1}{\beta})} e^{-\left| \frac{x-\zeta}{\theta} \right|^\beta} \tag{21.4}$$

A generalization of the normal distribution. Special cases include the normal, Laplace and uniform distributions.

$$\mathrm{ExpPower}(x \; ; \; \zeta, \theta, 1) = \mathrm{Laplace}(x \; ; \; \zeta, \theta)$$
$$\mathrm{ExpPower}(x \; ; \; \zeta, \theta, 2) = \mathrm{Normal}(x \; ; \; \zeta, \theta/\sqrt{2})$$
$$\lim_{\beta \to \infty} \mathrm{ExpPower}(x \; ; \; \zeta, \theta, \beta) = \mathrm{Uniform}(x \; ; \; \zeta - \theta, 2\theta)$$

Generalized K distribution [143]:

$$\text{GenK}(x \; ; s, \alpha_1, \alpha_2, \beta) = \frac{2|\beta|}{|s|\Gamma(\alpha_1)\Gamma(\alpha_2)} \left(\frac{x}{s}\right)^{\frac{1}{2}(\alpha_1+\alpha_2)\beta-1} K_{\alpha_1-\alpha_2}\left(2\left(\frac{x}{s}\right)^{\frac{\beta}{2}}\right)$$

(21.5)

$$x \geqslant 0, \alpha_1 > 0, \alpha_2 > 0$$

The Weibull transform of the K-distribution (21.8). Arises as the product of anchored Amoroso distributions with common Weibull parameters.

$$\text{GenK}(s_1 s_2, \alpha_1, \alpha_2, \beta) \sim \text{Amoroso}_1(0, s_1, \alpha_1, \beta) \, \text{Amoroso}_2(0, s_2, \alpha_2, \beta)$$

$$\sim s_1 \, \text{Gamma}_1(0, \alpha_1)^{\frac{1}{\beta}} \, s_2 \, \text{Gamma}_2(0, \alpha_2)^{\frac{1}{\beta}}$$

$$\sim s_1 s_2 \big(\text{Gamma}_1(1, \alpha_1) \, \text{Gamma}_2(1, \alpha_2)\big)^{\frac{1}{\beta}}$$

$$\sim s_1 s_2 \, \text{K}(1, \alpha_1, \alpha_2)^{\frac{1}{\beta}}$$

Generalized Pearson VII (generalized Cauchy, generalized-t) distribution [134, 144, 145, 95, 146, 147]:

$$\text{GenPearsonVII}(x \; ; a, s, m, \beta)$$

(21.6)

$$= \frac{\beta}{2|s|B(m - \frac{1}{\beta}, \frac{1}{\beta})} \left(1 + \left|\frac{x-a}{s}\right|^{\beta}\right)^{-m}$$

$$x, a, s, m, \beta \text{ in } \mathbb{R}$$

$$\beta > 0, \; m > 0, \; \beta m > 1$$

A generalization of the Pearson type VII distribution (9.1). Special cases include Pearson VII (9.1), Cauchy (9.6), Laha (20.18), Meridian (21.13) and exponential power (21.4) distributions,

$$\text{GenPearsonVII}(x \; ; a, s, m, 2) = \text{PearsonVII}(x \; ; a, s, m)$$
$$\text{GenPearsonVII}(x \; ; a, s, 1, 2) = \text{Cauchy}(x \; ; a, s)$$
$$\text{GenPearsonVII}(x \; ; a, s, 1, 4) = \text{Laha}(x \; ; a, s)$$
$$\text{GenPearsonVII}(x \; ; a, s, 2, 1) = \text{Meridian}(x \; ; a, s)$$
$$\lim_{m \to \infty} \text{GenPearsonVII}(x \; ; a, m^{1/\beta}\theta, m, \beta) = \text{ExpPower}(x \; ; a, \theta, \beta)$$

A related distribution is the half generalized Pearson VII (18.10), a special case of generalized beta prime (18.1).

Holtsmark distribution [148]:

$$\text{Holtsmark}(x \; ; \mu, c) = \text{Stable}(x \; ; \mu, c, \tfrac{3}{2}, 0) \qquad (21.7)$$

A symmetric stable distribution (21.20). Although the Holtsmark distribution cannot be expressed with elementary functions, it does have an analytic form in terms of hypergeometric functions [149].

$$\text{Holtsmark}(x \; ; \mu, c) = \tfrac{1}{\pi}\Gamma(\tfrac{5}{3}) \; {}_2F_3\left(\tfrac{5}{12}, \tfrac{11}{12}; \tfrac{1}{3}, \tfrac{1}{2}, \tfrac{5}{6}; -\tfrac{4}{729}\left(\tfrac{x-\mu}{c}\right)^6\right)$$
$$- \tfrac{1}{3\pi}\left(\tfrac{x-\mu}{c}\right)^2 \; {}_3F_4\left(\tfrac{3}{4}, 1, \tfrac{5}{4}; \tfrac{2}{3}, \tfrac{5}{6}, \tfrac{7}{6}, \tfrac{4}{3}; -\tfrac{4}{729}\left(\tfrac{x-\mu}{c}\right)^6\right)$$
$$+ \tfrac{7}{81\pi}\Gamma(\tfrac{4}{3})\left(\tfrac{x-\mu}{c}\right)^4 \; {}_2F_3\left(\tfrac{13}{12}, \tfrac{19}{12}; \tfrac{7}{6}, \tfrac{3}{2}, \tfrac{5}{3}; -\tfrac{4}{729}\left(\tfrac{x-\mu}{c}\right)^6\right)$$

K distribution [143, 150, 151, 152]:

$$K(x \; ; s, \alpha_1, \alpha_2) = \frac{2}{|s|\Gamma(\alpha_1)\Gamma(\alpha_2)}\left(\tfrac{x}{s}\right)^{\frac{1}{2}(\alpha_1+\alpha_2)-1} K_{\alpha_1-\alpha_2}\left(2\sqrt{\tfrac{x}{s}}\right) \qquad (21.8)$$
$$x \geqslant 0, \alpha_1 > 0, \alpha_2 > 0$$

Note that modified Bessel function of the second kind (p.177) is symmetric with respect to its argument, $K_\nu(+z) = K_\nu(-z)$. Thus the K-distribution is symmetric with respect to the two shape parameters, $K(x \; ; s, \alpha_1, \alpha_2) =$

$K(x ; s, \alpha_2, \alpha_1)$.

The K-distribution arises as the product of Gamma distributions [143, 151, 152].

$$K(s_1 s_2, \alpha_1, \alpha_2) \sim \mathrm{Gamma}_1(0, s_1, \alpha_1)\, \mathrm{Gamma}_2(0, s_2, \alpha_2)$$

The K-distribution has applications to radar scattering [150, 151] and superstatistical thermodynamics [153, Eq. 21].

Irwin-Hall (uniform sum) distribution [154, 155, 3]:

$$\mathrm{IrwinHall}(x ; n) = \frac{1}{2(n-1)!} \sum_{k=0}^{n} (-1)^k \binom{n}{k} (x-k)^{n-1} \operatorname{sgn}(x-k) \quad (21.9)$$

The sum of n independent standard uniform variates.

$$\mathrm{IrwinHall}(n) \sim \sum_{i=1}^{n} \mathrm{Uniform}_i(0,1)$$

Related to the Bates distribution (21.1). For $n = 1$ we recover the uniform distribution (1.1), and with $n = 2$ the triangular distribution (21.22).

Johnson S_U distributions [156, 2]:

$$\mathrm{JohnsonSU}(x ; \mu, \sigma, \gamma, \delta) = \frac{\delta}{\lambda\sqrt{2\pi}} \frac{1}{\sqrt{1 + \left(\frac{x-\xi}{\lambda}\right)^2}} e^{-\frac{1}{2}\left(\gamma + \delta \sinh^{-1}\left(\frac{x-\xi}{\lambda}\right)\right)^2}$$

$$(21.10)$$

Johnson's distributions are transforms of the normal distribution,

$$\mathrm{Johnson}_g(\mu, \sigma, \gamma, \delta) \sim \sigma g\left(\frac{\mathrm{StdNormal}() - \gamma}{\delta}\right) + \mu$$

Where for Johnson S_U the function is $g(x) = \sinh(x)$. For Johnson S_B the function is $g(x) = 1/(1 + \exp(x))$, for Johnson S_L, $g(x) = \exp(x)$) (i.e. log-normal), and for Johnson S_N the function is constant, recapitulating the normal distribution.

Landau distribution [157]:

$$\text{Landau}(x \; ; \mu, c) = \text{Stable}(x \; ; \mu, c, 1, 1) \qquad (21.11)$$

A stable distribution (21.20). Describes the average energy loss of a charged particles traveling through a thin layer of matter [157].

Log-Cauchy distribution [158]:

$$\text{LogCauchy}(x \; ; a, s, \beta) = \frac{|\beta|}{|s|\pi} \left(\frac{x-a}{s}\right)^{-1} \frac{1}{1 + \left(\ln\left(\frac{x-a}{s}\right)^{\beta}\right)^2} \qquad (21.12)$$

A log-stable distribution with very heavy tails. The anti-log transform of the Cauchy distribution (9.6).

$$\text{LogCauchy}(0, s, \beta) \sim \exp\left(-\text{Cauchy}(-\ln s, \tfrac{1}{\beta})\right)$$

Meridian distribution [147, Eq. 18] :

$$\text{Meridian}(x \; ; a, s) = \frac{1}{2|s|} \frac{1}{\left(1 + |\frac{x-a}{s}|\right)^2} \qquad (21.13)$$

The Laplace ratio distribution [147].

$$\text{Meridian}(x \; ; 0, \tfrac{s_1}{s_2}) \sim \frac{\text{Laplace}_1(0, s_1)}{\text{Laplace}_2(0, s_2)}$$

A special case of the generalized Pearson VII distribution (21.6).

Noncentral chi (Noncentral χ) distribution [33, 3]:

$$\text{NoncentralChi}(x ; k, \lambda) = \lambda e^{-\frac{1}{2}(x^2 + \lambda^2)} \left(\frac{x}{\lambda}\right)^{\frac{k}{2}} I_{\frac{k}{2} - 1}(\lambda x) \qquad (21.14)$$

$$k, \lambda, x \text{ in } \mathbb{R}, > 0$$

Here, $I_\nu(z)$ is a modified Bessel function of the first kind (p.177). A generalization of the chi distribution (11.8).

$$\text{NoncentralChi}(k, \lambda) \sim \sqrt{\text{NoncentralChiSqr}(k, \lambda)}$$

Noncentral chi-square (Noncentral χ^2, χ'^2) distribution [33, 3]:

$$\text{NoncentralChiSqr}(x ; k, \lambda) = \frac{1}{2} e^{-(x+\lambda)/2} \left(\frac{x}{\lambda}\right)^{\frac{k}{4} - \frac{1}{2}} I_{\frac{k}{2} - 1}(\sqrt{\lambda x}) \qquad (21.15)$$

$$k, \lambda, x \text{ in } \mathbb{R}, > 0$$

Here, $I_\nu(z)$ is a modified Bessel function of the first kind (p.177). A generalization of the chi-square distribution. The distribution of the sum of k squared, independent, normal random variables with means μ_i and standard deviations σ_i,

$$\text{NoncentralChiSqr}(k, \lambda) \sim \sum_{i=1}^{k} \left(\frac{1}{\sigma_i} \text{Normal}_i(\mu_i, \sigma_i)\right)^2$$

where the non-centrality parameter $\lambda = \sum_{i=1}^{k} (\mu_i / \sigma_i)^2$.

Noncentral F distribution [33, 3]:

$$\text{NoncentralF}(k_1, k_2, \lambda_1, \lambda_2) \sim \frac{\text{NoncentralChiSqr}_1(k_1, \lambda_1)/k_1}{\text{NoncentralChiSqr}_2(k_2, \lambda_2)/k_2}$$

$$\text{for } k_1, k_2, \lambda_1, \lambda_2 > 0$$

$$\text{support } x > 0 \qquad (21.16)$$

The ratio distribution of non-central chi square distributions. If both centrality parameters λ_1, λ_2 are non zero, then we have a **doubly non-central F distribution**; if one is zero then we have a **singly non-central F distribution**; and if both are zero we recover the standard F distribution (13.3).

Pseudo-Voigt distribution [159]:

$$\text{PseudoVoigt}(x \; ; \; a, \sigma, s, \eta) = (1 - \eta) \, \text{Normal}(x \; ; \; a, \sigma) + \eta \, \text{Cauchy}(x \; ; \; a, s)$$
$$\text{for } 0 \leqslant \eta \leqslant 1 \tag{21.17}$$

A linear mixture of Cauchy (Lorentzian) and normal distributions. Used as a more analytically tractable approximation to the Voigt distribution (21.24).

Rice (Rician, Rayleigh-Rice, generalized Rayleigh) distribution [160, 161]:

$$\text{Rice}(x \; ; \; \nu, \sigma) = \frac{x}{\sigma^2} \exp\left(-\frac{x^2 + \nu^2}{2\sigma^2}\right) I_0\left(\frac{x|\nu|}{\sigma^2}\right) \tag{21.18}$$
$$x > 0$$

Here, $I_0(z)$ is a modified Bessel function of the first kind (p.177).

The absolute value of a circular bivariate normal distribution, with non-zero mean,

$$\text{Rice}(\nu, \sigma) \sim \sqrt{\text{Normal}_1^2(\nu \cos \theta, \sigma) + \text{Normal}_2^2(\nu \sin \theta, \sigma)}$$

thus directly related to a special case of the noncentral chi-square distribution (21.15).

$$\text{Rice}(\nu, 1)^2 \sim \text{NoncentralChiSqr}(2, \nu^2)$$

Slash distribution [162, 2]:

$$\text{Slash}(x) = \frac{\text{StdNormal}(x) - \text{StdNormal}(x)}{x^2} \qquad (21.19)$$

The standard normal – standard uniform ratio distribution,

$$\text{Slash}() \sim \frac{\text{StdNormal}()}{\text{StdUniform}()}$$

Note that $\lim_{x \to 0} \text{Slash}(x) = 1/\sqrt{8\pi}$.

Stable (Lévy skew alpha-stable, Lévy stable) distribution [163]: The PDF of the stable distribution does not have a closed form in general. Instead, the stable distribution can be defined via the characteristic function

$$\text{StableCF}(t \,;\, \mu, c, \alpha, \beta) = \exp\big(it\mu - |ct|^\alpha(1 - i\beta\,\text{sgn}(t)\Phi(\alpha))\big) \qquad (21.20)$$

where $\Phi(\alpha) = \tan(\pi\alpha/2)$ if $\alpha \neq 1$, else $\Phi(1) = -(2/\pi)\log|t|$. Location parameter μ, scale c, and two shape parameters, the index of stability or characteristic exponent $\alpha \in (0, 2]$ and a skewness parameter $\beta \in [-1, 1]$. This distribution is continuous and unimodal [164], symmetric if $\beta = 0$ (**Lévy symmetric alpha-stable**), and indefinite support, unless $\beta = \pm 1$ and $0 < \alpha \leqslant 1$, in which case the support is semi-infinite. If c or α is zero, the distribution limits to the degenerate distribution, (§1). Non-normal stable distributions ($\alpha < 2$) are called **stable Paretian distributions**, since they all have long, Pareto tails.

A distribution is stable if it is closed under scaling and addition,

$$a_1 \,\text{Stable}_1(\mu, c, \alpha, \beta) + a_2 \,\text{Stable}_2(\mu, c, \alpha, \beta) \sim a_3 \,\text{Stable}_3(\mu, c, \alpha, \beta) + b$$

for real constants a_1, a_2, a_3, b. The anti-log transform of a stable distribution is log-stable: it is stable under multiplication instead of addition.

There are three special cases of the stable distribution where the probability density functions can be expressed with elementary functions: The normal (4.1), Cauchy (9.6), and Lévy (11.15) distributions, all of which are simple.

Table 21.1: Special cases of the stable family

(21.20)	stable	μ	c	α	β
(9.6)	Cauchy	.	.	1	0
(21.7)	Holtsmark	.	.	$\frac{3}{2}$	0
(4.1)	normal	.	.	2	0
(11.15)	Lévy	.	.	$\frac{1}{2}$	1
(21.11)	Landau	.	.	1	1

Suzuki distribution [165]. A compounded mixture of Rayleigh and log-normal distributions

$$\text{Suzuki}(\vartheta, \sigma) \sim \text{Rayleigh}(\sigma') \underset{\sigma'}{\wedge} \text{LogNormal}(0, \vartheta, \sigma) \tag{21.21}$$

Introduced to model radio propagation in cluttered urban environments.

Triangular (tine) distribution [68]:

$$\text{Triangular}(x \; ; \; a, b, c) = \begin{cases} \frac{2(x-a)}{(b-a)(c-a)} & a \leqslant x \leqslant c \\ \frac{2(b-x)}{(b-a)(b-c)} & c \leqslant x \leqslant b \end{cases} \tag{21.22}$$

Support $x \in [a, b]$ and mode c. The wedge distribution (5.4) is a special case.

Uniform difference distribution [49]:

$$\text{UniformDiff}(x) = \begin{cases} (1+x) & -1 \geqslant x \geqslant 0 \\ (1-x) & 0 \geqslant x \geqslant 1 \end{cases} \tag{21.23}$$

$$= \text{Triangular}(x \; ; -1, 1, 0)$$

The difference of two independent standard uniform distributions (1.2).

Voigt (Voigt profile, Voigtian) distribution [166]:

$$\text{Voigt}(a, \sigma, s) = \text{Normal}(0, \sigma) + \text{Cauchy}(a, s) \qquad (21.24)$$

The convolution of a Cauchy (Lorentzian) distribution with a normal distribution. Models the broadening of spectral lines in spectroscopy [166]. See also Pseudo Voigt distribution (21.17).

A NOTATION AND NOMENCLATURE

Notation

We write $\mathrm{Amoroso}(x \; ; \; a, \theta, \alpha, \beta)$ for a density function, $\mathrm{AmorosoCDF}(x \; ; \; a, \theta, \alpha, \beta)$ for the cumulative distribution function, $\mathrm{Amoroso}(a, \theta, \alpha, \beta)$ for the corresponding random variable, and $X \sim \mathrm{Amoroso}(a, \theta, \alpha, \beta)$ to indicate that two random variables have the same probability distribution [65]. The semicolon, which we verbalize as "given" or "parameterized by", separates the arguments from the parameters.

parameter	type	notes	
a	location	power-function	
b	location	arcsine, $b = a + s$	
ζ	location	exponential	eta
μ	location	normal	mu
ν	location	gamma-exponential	nu
ζ	location	beta-exponential	zeta
s	scale	power function	
λ	scale	exponential	lambda
σ	scale	normal	sigma
$\vartheta^\dagger$	scale	log-normal	theta
θ	scale	Amoroso	theta
ω	scale	gen. Fisher Tippett	omega
β	power	power function	beta
α	shape	> 0, beta and beta prime families	alpha
γ	shape	> 0, beta and beta prime families	gamma
n	shape	integer > 0, number of samples or events	
k	shape	integer > 0, degrees of freedom	
m	shape	$> \frac{1}{2}$, Pearson IV	
ν	shape	> 0, Pearson IV	

† A curly theta, or "vartheta".

Throughout I have endeavored to use consistent parameterization, both within families, and between subfamilies and superfamilies. For instance, β is always the Weibull power parameter. Location (or translation) parameters: a, b, ν, μ. Scale parameters: s, θ, σ. Shape parameters: α, γ, m, ν. All parameters are real and the shape parameters α, γ and m are positive.

The negation of a standard parameter is indicated by a bar, e.g. $\beta = -\bar{\beta}$. For clarity we use a dot '.' in tables of special cases to indicate repetition of the base distribution's parameters.

Nomenclature

interesting   Informally, an "interesting distribution" is one that has acquired a name, which generally indicates that the distribution is the solution to one or more interesting problems.

generalized-X   The only consistent meaning is that distribution "X" is a special case of the distribution "generalized-X". In practice, often means "add another parameter". We use alternative nomenclature whenever practical, and generally reserve "generalized" for the power (Weibull) transformed distribution.

standard-X   The distribution "X" with the location parameter set to 0 and scale to 1. Not to be confused with *standardized* which generally indicates zero mean and unit variance.

shifted-X   (or translated-X) A distribution with an additional location parameter.

anchored-X   (or ballasted-X) A distribution with a fixed location (typically with a lower bound set to zero).

scaled-X   (or scale-X) A distribution with an additional scale parameter.

inverse-X   (Occasionally inverted-X, reciprocal-X, or negative-X) Generally labels the transformed distribution with $x \to \frac{1}{x}$, or more generally the distribution with the Weibull shape parameter negated, $\beta \to -\beta$. An exception is the inverse Gaussian distribution (20.3) [2].

log-X   Either the anti-logarithmic or logarithmic transform of the random variable X, i.e. either $\exp{-X()} \sim$ log-X() (e.g. log-normal) or $-\ln X() \sim$ log-X(). This ambiguity arises because although the second convention may seem more logical, the log-normal convention has historical precedence. Herein, we follow the log-normal convention.

X-exponential   The logarithmic transform of distribution X, i.e. $\ln X() \sim$ X-exponential(). This naming convention, which arises from the beta-exponential distribution (14.1), sidesteps the confusion surrounding the log-X naming convention.

reversed-X   (Occasionally negative-X) The scale is negated.

X of the Nth kind   See "X type N".

folded-X   The distribution of the absolute value of random variable X.

beta-X   A distribution formed by inserting the cumulative distribution function of X into the CDF of the standard beta distribution (12.2). Distributions of this form arise naturally in the study of order statistics (§C).

central-X   A distribution formed by inserting the cumulative distribution function of X into the CDF of the central beta distribution (12.5). Distributions of this form arise naturally in the study of median statistics (§C).[1]

B   PROPERTIES OF DISTRIBUTIONS

notation   The multi-letter, camel-cased function name, arguments and parameters used for the probability density of the family in this text.

probability density function (PDF)   The probability density $f_X(x)$ of a continuous random variable is the relative likelihood that the random variable will occur at a particular point. The probability to occur within a particular interval is given by the integral

$$P[a \leqslant X \leqslant b] = \int_a^b f_X(x)dx \,.$$

cumulative density function (CDF)   The probability that a random variable has a value equal or less than x, typically denoted by $F_X(x)$, and also called the distribution function for short.

$$F_X(x) = \int_{-\infty}^x f_X(z)dz$$

The probability density is equal to the derivative of the distribution function, assuming that the distribution function is continuous.

$$f_X(x) = \frac{d}{dx}F_X(x)$$

Negating a scale parameter gives a reversed distribution with the cumulative distribution function replaced by the complimentary cumulative distribution function (CCDF = 1 − CDF).

complimentary cumulative density function (CCDF)   (survival function, reliability function)   One minus the cumulative distribution function, $1 - F_X(x)$. The probability that a random variable has a value greater than x. In lifetime analysis the complimentary cumulative distribution function is also called the survival function or reliability function.

support   The support of a probability density function are the set of values that have non-zero probability. The compliment of the support has zero probability. The range (or image) of a random variable (the set of values that can be generated) is the support of the corresponding probability density.

mode The point where the distribution reaches its maximum value. An anti-mode is the point where the distribution reaches its minimum value. A distribution is called unimodal if there is only one local extremum away from the boundaries of the distribution. In other words, the distribution can have one mode ⌢ or one anti-mode ⌣, or be monotonically increasing / or decreasing \.

mean The expectation value of the random variable.

$$\mathbb{E}[X] = \int x\, f_X(x)\, dx$$

Not all interesting distributions have finite means, notably the Cauchy family (9.6). Often denoted by the symbol μ.

variance The variance measures the spread of a distribution.

$$\mathrm{var}[X] = \mathbb{E}\big[(X - \mathbb{E}[X])^2\big] = \mathbb{E}\big[X^2\big] - \mathbb{E}\big[X\big]^2$$

The variance is also know as the second central moment, or second cumulant, and commonly denoted by the symbol σ. The standard deviation is the square root of the variance.

central moment

$$\mu_n[X] = \mathbb{E}\big[(X - \mathbb{E}[X])^n\big] \tag{2.1}$$

The nth moment about the mean. The first central moment is zero, and the second is the variance.

skew A distribution is skewed if it is not symmetric. A positively skewed distribution tends to have a majority of the probability density above the mean; a negatively skewed distribution tends to have a majority of density below the mean.

The standard measure of skew is the third cumulant (third central moment) normalized by the $\frac{3}{2}$ power of the second cumulant.

$$\mathrm{skew}[X] = \mathbb{E}\left[\left(\frac{X - \mathbb{E}[X]}{\sigma[X]}\right)^3\right] = \frac{\kappa_3}{\kappa_2^{\frac{3}{2}}}$$

kurtosis   Kurtosis measures the spread of a distribution. The normal distribution has zero excess kurtosis. A positive kurtosis distribution longer tails, while a negative kurtosis distribution has shorter tails.

The standard measure of kurtosis is the forth cumulant normalized by the square of the second cumulant.

$$\text{ExKurtosis}[X] = \frac{\kappa_4}{\kappa_2{}^2}$$

This measure is called the excess kurtosis to distinguish it from an older definition of kurtosis that used the forth central moment μ_4 instead of the forth cumulant. (Note that $\frac{\kappa_4}{\kappa_2{}^2} = \frac{\mu_4}{\kappa_2{}^2} - 3$).

entropy   The differential (or continuous) entropy of a continuous probability distribution is

$$\text{entropy}[X] = -\int f(x) \ln f(x) \, dx$$

Note that unlike the entropy of a discrete variable, the differential entropy is not invariant under a change of variables, and can be negative.

moment generating function (MGF)   The expectation

$$\text{MGF}_X(t) = \mathbb{E}[e^{tX}] \,.$$

The nth derivative of the moment generating function, evaluated at 0, is equal to the nth moment of the distribution.

$$\frac{d^n}{dt^n} \text{MGF}_X(t)\Big|_0 = \mathbb{E}[X^n]$$

If two random variables have identical moment generating functions, then they have identical probability densities.

cumulant generating function (CGF)   The logarithm of the moment generating function.

$$\text{CGF}_X(t) = \ln \mathbb{E}[e^{tX}]$$

Note that some authors define the cumulant generating function as the logarithm of the characteristic function.

The nth derivative of the cumulant generating function, evaluated at 0, is equal to the nth cumulant of the distribution.

$$\frac{d^n}{dt^n} \mathrm{CGF}_X(t)\Big|_0 = \kappa_n(X) \tag{2.2}$$

The nth cumulant is a function of the first n moments of the distribution, and the second and third are equal to the second and third central moments.

$$\kappa_1 = \mathbb{E}[X]$$
$$\kappa_2 = \mathbb{E}\big[(X - \mathbb{E}[X])^2\big]$$
$$\kappa_3 = \mathbb{E}\big[(X - \mathbb{E}[X])^3\big]$$
$$\kappa_4 = \mathbb{E}\big[(X - \mathbb{E}[X])^4\big] - 3\,\mathbb{E}\big[(X - \mathbb{E}[X])^2\big]$$

The cumulant expansion, if it exists, either terminates at second order (normal distribution), or continues to infinite order.

Cumulants are often more useful than central moments, since cumulants are additive under summation of independent random variables.

$$\mathrm{CGF}_{X+Y}(t) = \mathrm{CGF}_X(t) + \mathrm{CGF}_Y(t)$$

characteristic function (CF) Neither the moment nor cumulant generating functions need exist for a given distribution. An alternative that always exists is the characteristic function

$$\phi_X(t) = \mathbb{E}[e^{itX}]\,,$$

essentially the Fourier transform of the probability density function. The characteristic function for a sum of independent random variables is the product of the respective characteristic functions.

$$\phi_{X+Y}(t) = \phi_X(t)\,\phi_Y(t)$$

More generally, the characteristic function of any linear sum of independent random variables is

$$\phi_Z(t) = \prod_i \phi_{X_i}(c_i t), \quad Z = \sum_i c_i X_i\,.$$

quantile function   The inverse of the cumulative distribution function, typically denoted $F^{-1}(p)$ (or occasionally $Q(p)$). The median is the middle value of the inverse cumulative distribution function.

$$\mathrm{median}[X] = F_X^{-1}(\tfrac{1}{2})$$

Half the probability density is above the median, half below. The quantile and median rarely have simple forms.

hazard function   The ratio of the probability density function to the complimentary cumulative distribution function

$$\mathrm{hazard}_X(x) = \frac{f_X(x)}{1 - F_X(x)}$$

C   Order statistics

Order statistics

Order statistics [167]: If we draw $m+n-1$ independent samples from a distribution, then the distribution of the mth smallest value (or equivalently the nth largest) is

$$\text{OrderStatistic}_X(x \; ; m, n) = \frac{(m+n-1)!}{(m-1)!(n-1)!} \, F(x)^{m-1} \, f(x) \, (1 - F(x))^{n-1}$$

Here X is a random variable, $f(x)$ is the corresponding probability density and $F(x)$ is the cumulative distribution function. The first term is the number of ways to separate $m+n-1$ things into three groups containing 1, $m-1$, and $n-1$ things; the second is the probability of drawing $m-1$ samples smaller than the sample of interest; the third term is the distribution of the mth sample; and the fourth term is the probability of drawing $n-1$ larger samples. Note that the smallest value is obtained if $m = 1$, the largest value if $n = 1$, and the median value if $m = n$.

The cumulative distribution function (CDF) for order statistics can be written in terms of the regularized beta function, $I(p, q; z)$.

$$\text{OrderStatisticCDF}_X(x \; ; m, n) = I(m, n; F(x))$$

Conversely, if a CDF for a distribution has the form $I(m, n; F(x))$, then $F(x)$ is the cumulative distribution function of the corresponding ordering distribution. Since $I(\alpha, \gamma; x)$ is the CDF of the beta distribution (12.1), beta-generalized distributions of the form $I(\alpha, \gamma; F_X(x))$ (with arbitrary positive α and γ) are often referred to as 'beta-X' [168], e.g. the beta-exponential distribution (14.1).

The order statistic of the uniform distribution (1.1) is the beta distribution (12.1), that of the exponential distribution (2.1) is the beta-exponential distribution (14.1), and that of the power function distribution (5.1) is the generalized beta distribution (17.1).

$$\mathrm{OrderStatistic}_{\mathrm{Uniform}(a,s)}(x\ ;\ \alpha,\gamma) = \mathrm{Beta}(x\ ;\ a,s,\alpha,\gamma)$$

$$\mathrm{OrderStatistic}_{\mathrm{Exp}(\zeta,\lambda)}(x\ ;\ \gamma,\alpha) = \mathrm{BetaExp}(x\ ;\ \zeta,\lambda,\alpha,\gamma)$$

$$\mathrm{OrderStatistic}_{\mathrm{PowerFn}(a,s,\beta)}(x\ ;\ \alpha,\gamma) = \mathrm{GenBeta}(x\ ;\ a,s,\alpha,\gamma,\beta)$$

$$\mathrm{OrderStatistic}_{\mathrm{UniPrime}(a,s)}(x\ ;\ \alpha,\gamma) = \mathrm{BetaPrime}(x\ ;\ a,s,\alpha,\gamma)$$

$$\mathrm{OrderStatistic}_{\mathrm{Logistic}(\zeta,\lambda)}(x\ ;\ \gamma,\alpha) = \mathrm{BetaLogistic}(x\ ;\ \zeta,\lambda,\alpha,\gamma)$$

$$\mathrm{OrderStatistic}_{\mathrm{LogLogistic}(a,s,\beta)}(x\ ;\ \alpha,\gamma) = \mathrm{GenBetaPrime}(x\ ;\ a,s,\alpha,\gamma,\beta)$$

Extreme order statistics

In the limit that $n \gg m$ (or equivalently $m \gg n$) we obtain the distributions of *extreme order statistics*. Extreme order statistics depends only on the tail behavior of the sampled distribution; whether the tail is finite, exponential or power-law. This explains the central importance of the generalized beta distribution (17.1) to order statistics, since the power function distribution (5.1) displays all three classes of tail behavior, depending on the parameter β. Consequentially, the generalized beta distribution limits to the generalized Fisher-Tippett distribution (11.24), which is the parent of the other, specialized extreme order statistics. See also extreme order statistics, (§11).

Median statistics

If we draw N independent samples from a distribution (Where N is odd), then the distribution of the statistical median value is

$$\mathrm{MedianStatistic}_X(x\ ;\ N) = \mathrm{OrderStatistic}_X(x\ ;\ \tfrac{N+1}{2}, \tfrac{N+1}{2})$$

Notable examples of median statistic distributions include

$$\mathrm{MedianStatistics}_{\mathrm{Uniform}(a,s)}(x\ ;\ 2\alpha - 1) = \mathrm{CentralBeta}(x\ ;\ a+s, 2s, \alpha)$$

$$\mathrm{MedianStatistics}_{\mathrm{Logistic}(a,s)}(x\ ;\ 2\alpha - 1) = \mathrm{CentralLogistic}(x\ ;\ a,s,\alpha)$$

The median statistics of symmetric distributions are also symmetric.

Figure 39: Order Statistics

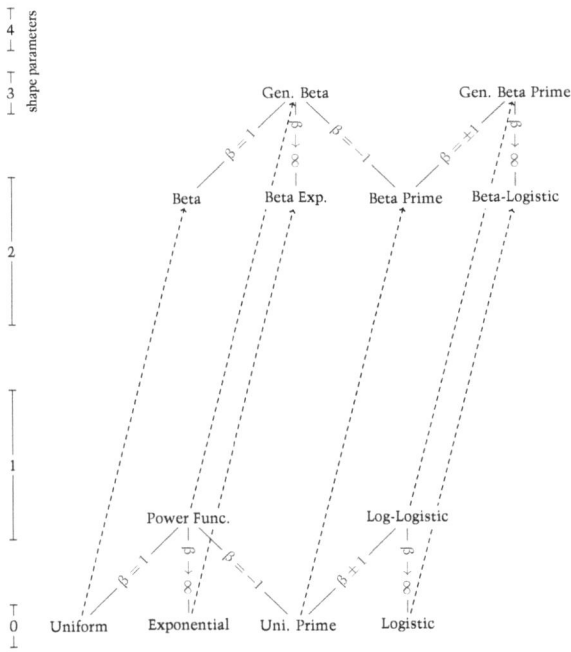

D Limits

Exponential function limit

A common and important limit is

$$\lim_{c \to +\infty} \left(1 + \frac{x}{c}\right)^{ac} = e^{ax} .$$

In particular, the X-exponential distributions are the exponential limit of Weibullized distributions.

$$\lim_{\beta \to \infty} f\left[\left(\frac{x-a}{s}\right)^\beta\right] = \lim_{\beta \to \infty} f\left[\left(1 - \frac{1}{\beta}\frac{x-\zeta}{\lambda}\right)^\beta\right] = f\left[e^{-\frac{x-\zeta}{\lambda}}\right]$$

$$(a = \zeta + \beta\lambda, \ s = -\beta\lambda)$$

$$\mathrm{Exp}(x \ ; \ a, \theta) = \lim_{\beta \to \infty} \mathrm{PowerFn}(x \ ; \ a + \beta\theta, -\beta\theta, \beta)$$

$$\mathrm{GammaExp}(x \ ; \ \nu, \lambda, \alpha) = \lim_{\beta \to \infty} \mathrm{Amoroso}(x \ ; \ \nu + \beta\lambda, -\beta\lambda, \alpha, \beta)$$

$$\mathrm{Gamma}(x \ ; \ a, s, \alpha) = \lim_{\beta \to \infty} \mathrm{UnitGamma}(x \ ; \ a + \beta s, -\beta s, \alpha, \beta)$$

$$\mathrm{BetaExp}(x \ ; \ \zeta, \lambda, \alpha, \gamma) = \lim_{\beta \to \infty} \mathrm{GenBeta}(x \ ; \ \zeta + \beta\lambda, -\beta\lambda, \alpha, \gamma, \beta)$$

$$\mathrm{BetaLogistic}(x \ ; \ \zeta, \lambda, \alpha, \gamma) = \lim_{\beta \to \infty} \mathrm{GenBetaPrime}(x \ ; \ \zeta + \beta\lambda, -\beta\lambda, \alpha, \gamma, \beta)$$

$$\mathrm{Normal}(x \ ; \ \mu, \sigma) = \lim_{\beta \to \infty} \mathrm{LogNormal}(x \ ; \ \mu + \beta\sigma, -\beta\sigma, \beta)$$

We can play the same trick with the γ shape parameter in the beta and beta prime families.

$$\lim_{\gamma \to \infty} f\left[\left(1 - \left(\frac{x-a}{s}\right)^\beta\right)^{\gamma-1}\right] = \lim_{\gamma \to \infty} f\left[\left(1 - \frac{1}{\gamma}\left(\frac{x-a}{\theta}\right)^\beta\right)^{\gamma-1}\right]$$

$$= f\left[e^{-\left(\frac{x-a}{\theta}\right)^\beta}\right] \qquad s = \theta\gamma^{\frac{1}{\beta}}$$

$$\mathrm{Amoroso}(x \ ; \ a, \theta, \alpha, \beta) = \lim_{\gamma \to \infty} \mathrm{GenBeta}(x \ ; \ a, \theta\gamma^{\frac{1}{\beta}}, \alpha, \gamma, \beta)$$

$$\mathrm{Gamma}(x \ ; \ a, \theta, \alpha) = \lim_{\gamma \to \infty} \mathrm{Beta}(x \ ; \ a, \theta\gamma, \alpha, \gamma)$$

$$\lim_{\gamma\to\infty} f\left[\left(1+\left(\frac{x-a}{s}\right)^\beta\right)^{-\alpha-\gamma}\right] = \lim_{\gamma\to\infty} f\left[\left(1+\frac{1}{\gamma}\left(\frac{x-a}{\theta}\right)^\beta\right)^{-\alpha-\gamma}\right]$$

$$= f\left[e^{-\left(\frac{x-a}{\theta}\right)^\beta}\right] \qquad s=\theta\gamma^{\frac{1}{\beta}}$$

$$\mathrm{Amoroso}(x \; ; \; a,\theta,\alpha,\beta) = \lim_{\gamma\to\infty} \mathrm{GenBetaPrime}(x \; ; \; a,\theta\gamma^{\frac{1}{\beta}},\alpha,\gamma,\beta)$$

$$\mathrm{Gamma}(x \; ; \; 0,\theta,\alpha) = \lim_{\gamma\to\infty} \mathrm{BetaPrime}(x \; ; \; 0,\theta\gamma,\alpha,\gamma)$$

$$\mathrm{InvGamma}(x \; ; \; \theta,\alpha) = \lim_{\gamma\to\infty} \mathrm{BetaPrime}(x \; ; \; 0,\theta/\gamma,\alpha,\gamma)$$

Essentially the same limit takes the beta-exponential and beta-logistic distributions to the Gamma-Exponential distribution.

$$\mathrm{GammaExp}(x \; ; \; \nu,\lambda,\alpha) = \lim_{\gamma\to\infty} \mathrm{BetaExp}(x \; ; \; \nu+\lambda/\ln\gamma,\lambda,\alpha,\gamma)$$

$$\mathrm{GammaExp}(x \; ; \; \nu,\lambda,\alpha) = \lim_{\gamma\to\infty} \mathrm{BetaLogistic}(x \; ; \; \nu+\lambda/\ln\gamma,\lambda,\alpha,\gamma)$$

$$\mathrm{Gumbel}(x \; ; \; \nu,\lambda) = \lim_{\gamma\to\infty} \mathrm{ExpExp}(x \; ; \; \nu+\lambda/\ln\gamma,\lambda,\gamma)$$

$$\mathrm{Gumbel}(x \; ; \; \nu,\lambda) = \lim_{\gamma\to\infty} \mathrm{BurrII}(x \; ; \; \nu+\lambda/\ln\gamma,\lambda,\gamma)$$

Logarithmic function limit

$$\lim_{c\to 0} \frac{x^c-1}{c} = \ln x$$

$$\mathrm{UnitGamma}(x \; ; \; a,s,\gamma,\beta) = \lim_{\alpha\to\infty} \mathrm{GenBeta}(x \; ; \; a,s,\alpha,\gamma,\beta/\alpha)$$

Gaussian function limit

$$\lim_{c\to\infty} e^{-z\sqrt{c}}\left(1+\frac{z}{\sqrt{c}}\right)^c = e^{-\frac{1}{2}z^2}$$

$$\text{LogNormal}(x \; ; \; a, \vartheta, \sigma) = \lim_{\gamma \to \infty} \text{UnitGamma}(x \; ; \; a, \vartheta e^{\sigma\sqrt{\gamma}}, \alpha, \tfrac{\sqrt{\gamma}}{\sigma})$$

$$\text{Normal}(x \; ; \; \mu, \sigma) = \lim_{\alpha \to \infty} \text{Gamma}(x \; ; \; \mu - \sigma\sqrt{\alpha}, \tfrac{\sigma}{\sqrt{\alpha}}, \alpha)$$

$$\text{Normal}(x \; ; \; \mu, \sigma) = \lim_{\alpha \to \infty} \text{InvGamma}(x \; ; \; \mu - \sigma\sqrt{\alpha}, \sigma\alpha^{\frac{3}{2}}, \alpha)$$

$$\lim_{c \to \infty} e^{c + c\frac{z}{\sqrt{c}} - ce^{\frac{z}{\sqrt{c}}}} = e^{-\frac{z^2}{2}}$$

$$\text{LogNormal}(x \; ; \; a, \vartheta, \sigma) = \lim_{\alpha \to \infty} \text{Amoroso}(x \; ; \; a, \vartheta\alpha^{-\sigma\sqrt{\alpha}}, \alpha, \tfrac{1}{\sigma\sqrt{\alpha}})$$

$$\text{Normal}(x \; ; \; \mu, \sigma) = \lim_{\alpha \to \infty} \text{GammaExp}(x \; ; \; \mu + \sigma\sqrt{\alpha}\ln\alpha, \sigma\sqrt{\alpha}, \alpha)$$

Miscellaneous limits

$$\text{InvGamma}(x \; ; \; \theta, \alpha) = \lim_{\nu \to \infty} \text{PearsonIV}(x \; ; \; 0, -\tfrac{\theta}{2\nu}, \tfrac{\alpha+1}{2}, \nu)$$

See (§16)

$$\text{Normal}(x \; ; \; \mu, \sigma) = \lim_{m \to \infty} \text{PearsonVII}(x \; ; \; \mu, \sigma\sqrt{2m}, m)$$

$$\text{Normal}(x \; ; \; \mu, \sigma) = \lim_{\alpha \to \infty} \text{CentralBeta}(x \; ; \; \mu, \sigma\sqrt{8\alpha}, \alpha)$$

$$\text{Laplace}(x \; ; \; \eta, \theta) = \lim_{\alpha \to 0} \text{BetaLogistic}(x \; ; \; \eta, \theta\alpha, \alpha, \alpha)$$

Figure 40: Limits and special cases of principal distributions

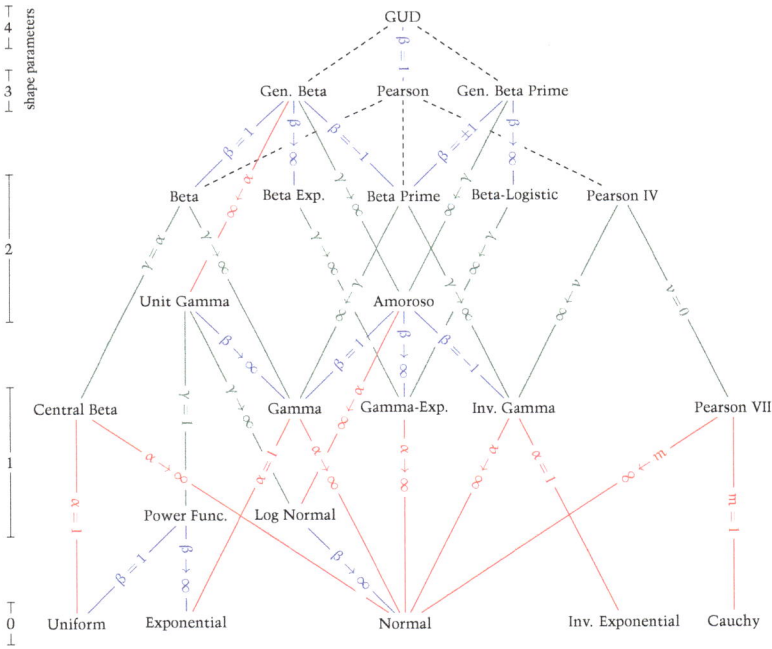

E   Algebra of Random Variables

Various operations can be applied to combine or transform random variables, providing a rich tapestry of interrelations between different distributions [49, 42].

Transformations

Given a continuous random variable X, with distribution function F_X and density f_X, and a monotonic function $h(x)$ (either strictly increasing or strictly decreasing) on the range of X, we can create a new random variable Y,

$$Y \sim h(X)$$

$$F_Y(y) = \begin{cases} F_X(h^{-1}(y)) & h(x) \text{ is increasing function} \\ 1 - F_X(h^{-1}(y)) & h(x) \text{ is decreasing function} \end{cases}$$

$$f_Y(y) = \left| \frac{d}{dy} h^{-1}(y) \right| f_X(h^{-1}(y))$$

In the last line above, the prefactor is the *Jacobian* of the transformation.
   For h (And h^{-1}) increasing we have

$$F_Y(y) = P(Y \leqslant y) = P(h(X) \leqslant y) = P(X \leqslant h^{-1}(y)) = F_X(h^{-1}(y))$$

and decreasing

$$F_Y(y) = P(Y \leqslant y) = P(h(X) \leqslant y) = P(X \geqslant h^{-1}(y)) = 1 - F_X(h^{-1}(y)) .$$

Linear transformation

$$h(x) = a + sx$$

A linear transform creates a *location-scale family* of distributions.

Weibull transformation

$$h(x) = a + sx^{\frac{1}{\beta}}$$

The Weibull transform only applies to distributions with non-negative support.

$$\text{PowerFn}(a, s, \beta) \sim a + s \ \text{StdUniform}()^{\frac{1}{\beta}}$$

$$\text{Weibull}(a, \theta, \beta) \sim a + \theta \ \text{StdExp}()^{\frac{1}{\beta}}$$

$$\text{LogNormal}(a, \vartheta, \beta) \sim a + \vartheta \ \text{StdLogNormal}()^{\frac{1}{\beta}}$$

$$\text{Amoroso}(a, \theta, \alpha, \beta) \sim a + \theta \ \text{StdGamma}(\alpha)^{\frac{1}{\beta}}$$

$$\text{GenBeta}(a, s, \alpha, \gamma, \beta) \sim a + s \ \text{StdBeta}(\alpha, \gamma)^{\frac{1}{\beta}}$$

$$\text{GenBetaPrime}(a, s, \alpha, \gamma, \beta) \sim a + s \ \text{StdBetaPrime}(\alpha, \gamma)^{\frac{1}{\beta}}$$

The Weibull transform is increasing if $\frac{s}{\beta} > 0$, and decreasing if $\frac{s}{\beta} < 0$.

Inverse (reciprocal) transformation

$$h(x) = x^{-1}$$

The Weibull transform with $a = 0$, $s = 1$, and $\beta = -1$.

$$\text{Gamma}(0, 1, \alpha) \sim \text{InvGamma}(0, 1, \alpha)^{-1}$$
$$\text{Exp}(0, 1) \sim \text{InvExp}(0, 1)^{-1}$$
$$\text{Cauchy}(0, 1) \sim \text{Cauchy}(0, 1)^{-1}$$

Log and anti-log transformations

$$h(x) = -\ln(x) \qquad h(x) = \exp(-x)$$

The log and anti-log transforms are inverses of one another. See p.154 for a discussion of transformed distribution naming conventions.

$$\text{StdUniform}() \sim \exp\big(-\text{StdExp}()\big)$$
$$\text{StdLogNormal}() \sim \exp\big(-\text{StdNormal}()\big)$$
$$\text{StdGamma}(\alpha) \sim \exp\big(-\text{StdGammaExp}(\alpha)\big)$$
$$\text{StdBeta}(\alpha, \gamma) \sim \exp\big(-\text{StdBetaExp}(\alpha, \gamma)\big)$$
$$\text{StdBetaPrime}(\alpha, \gamma) \sim \exp\big(-\text{StdBetaLogistic}(\alpha, \gamma)\big)$$

The anti-log transform converts a location parameter into a scale parameter, and a scale parameter into a Weibull shape parameter.

$$\text{PowerFn}(0, s, \beta) \sim \exp\left(-\text{Exp}(-\ln s, \tfrac{1}{\beta})\right)$$
$$\text{LogLogistic}(0, s, \beta) \sim \exp\left(-\text{Logistic}(-\ln s, \tfrac{1}{\beta})\right)$$
$$\text{FisherTippett}(0, s, \beta) \sim \exp\left(-\text{Gumbel}(-\ln s, \tfrac{1}{\beta})\right)$$
$$\text{Amoroso}(0, s, \alpha, \beta) \sim \exp\left(-\text{GammaExp}(-\ln s, \tfrac{1}{\beta}, \alpha)\right)$$
$$\text{LogNormal}(0, \vartheta, \beta) \sim \exp\left(-\text{Normal}(-\ln \vartheta, \tfrac{1}{\beta})\right)$$
$$\text{UnitGamma}(0, s, \alpha, \beta) \sim \exp\left(-\text{Gamma}(-\ln s, \tfrac{1}{\beta}, \alpha)\right)$$
$$\text{GenBeta}(0, s, \alpha, \gamma, \beta) \sim \exp\left(-\text{BetaExp}(-\ln s, \tfrac{1}{\beta}, \alpha, \gamma)\right)$$
$$\text{GenBetaPrime}(0, s, \alpha, \gamma, \beta) \sim \exp\left(-\text{BetaLogistic}(-\ln s, \tfrac{1}{\beta}, \alpha, \gamma)\right)$$

Prime transformation [1]

$$\text{prime}(x) = \frac{1}{\frac{1}{x} - 1}, \quad \text{prime}^{-1}(y) = \frac{1}{\frac{1}{y} + 1}$$

This transformation relates the beta and beta-prime distributions.

$$\text{StdUniPrime}() \sim \text{prime}(\text{StdUniform}())$$
$$\text{StdBetaPrime}(\alpha, \gamma) \sim \text{prime}(\text{StdBeta}(\alpha, \gamma))$$

Combinations

Sum The sum of two random variables is

$$Z \sim X + Y$$

The resultant probability distribution function is the convolution of the component distribution functions.

$$f_Z(z) = (f_X * f_Y)(z) = \int_{-\infty}^{+\infty} f_X(x)\, f_Y(z - x)\, dx$$

The characteristic function for a sum of independent random variables is the product of the respective characteristic functions (p159).

$$\phi_{X+Y}(t) = \phi_X(t)\phi_Y(t)$$

Examples:

$$\text{Normal}_1(\mu_1, \sigma_1) + \text{Normal}_2(\mu_2, \sigma_2) \sim \text{Normal}_3(\mu_1 + \mu_2, \sqrt{\sigma_1^2 + \sigma_2^2})$$
$$\text{Exp}_1(a_1, \theta) + \text{Exp}(a_2, \theta) \sim \text{Gamma}(a_1, a_2, \theta, 2)$$
$$\text{Gamma}_1(a_1, \theta, \alpha_1) + \text{Gamma}_2(a_2, \theta, \alpha_2) \sim \text{Gamma}_3(a_1 + a_2, \theta, \alpha_1 + \alpha_2)$$

Stable distributions (21.20) are those that are invariant under summation, changing only location and scale.

Difference   The difference of two random variables.

$$Z \sim X - Y$$

$$\phi_{X-Y}(t) = \phi_X(t)\phi_Y(-t)$$

Examples:

$$\text{UniformDiff}(x) \sim \text{StdUniform}_1(x) - \text{StdUniform}_2(x)$$
$$\text{BetaLogistic}(x \; ; \zeta_1 - \zeta_2, \lambda, \alpha, \gamma) \sim \text{GammaExp}_1(x \; ; \zeta_1, \lambda, \alpha)$$
$$- \text{GammaExp}_2(x \; ; \zeta_2, \lambda, \gamma)$$

Product   A *product distribution* is the product of two independent random variables.

$$Z \sim XY$$

The probability distribution of Z is

$$f_Z(z) = \int f_X(x)\, f_Y\!\left(\frac{z}{x}\right) \frac{1}{|x|}\,dx$$

Examples:

$$\prod_{i=1}^{n} \text{Uniform}_i(0, 1) \sim \text{UniformProduct}(n)$$

$$\prod_{i=1}^{n} \text{PowerFn}_i(0, s_i, \beta) \sim \text{UnitGamma}(0, \prod_{i=1}^{n} s_i, n, \beta)$$

$$\prod_{i=1}^{n} \text{UnitGamma}_i(0, s_i, \alpha_i, \beta) \sim \text{UnitGamma}(0, \prod_{i=1}^{n} s_i, \sum_{i=1}^{n} \alpha_i, \beta)$$

$$\prod_{i=1}^{n} \text{LogNormal}_i(0, \vartheta_i, \beta_i) \sim \text{LogNormal}_i(0, \prod_{i=1}^{n} \vartheta_i, (\sum_{i=0}^{n} \beta_i^{-2})^{-\frac{1}{2}})$$

Ratio   The ratio (or quotient) distribution is the ratio of two random variables.

$$R \sim \frac{X}{Y}$$

Examples:

$$\text{StdBetaPrime}(\alpha, \gamma) \sim \frac{\text{StdGamma}_1(\alpha)}{\text{StdGamma}_2(\gamma)}$$

$$\text{StdCauchy}() \sim \frac{\text{StdNormal}_1()}{\text{StdNormal}_2()}$$

Mixture   A mixture (or compound) of two distributions is formed by selecting a parameter of one distribution from the probability distribution of the other.

$$Z(x\,;\,\alpha) = \int X(x\,;\,\beta) Y(\beta\,;\,\alpha)\,d\beta$$

For random variables this can be notated as

$$Z(\alpha) \sim X\big(Y(\alpha)\big)$$
$$\text{or} \quad Z(\alpha) \sim X(\beta) \bigwedge_{\beta} Y(\alpha)\,.$$

The name 'X-Y' is sometimes assigned to a compound of distributions 'X' and 'Y', although this is ambiguous when there are multiple parameters that could be compounded.

Transmutations

Fold Folded distributions arise when only magnitude, and not the sign, of a random variable is observed.

$$\text{Folded}_X(\zeta) \sim |X - \zeta|$$

An important example is the **folded normal** distribution

$$\text{FoldedNormal}(x \; ; \mu, \sigma)$$
$$= \tfrac{1}{2}\text{Normal}(x \; ; +\mu, \sigma) + \tfrac{1}{2}\text{Normal}(x \; ; -\mu, \sigma)$$
$$\text{for} \quad x, \mu, \sigma \text{ in } \mathbb{R}, x \geqslant 0$$

If we fold about the center of a symmetric distribution we obtain a 'halved' distribution. Examples already encountered are the half normal (11.7), half-Pearson type VII (18.8), and half-Cauchy (18.9) distributions. A halved Laplace (3.1) distribution is exponential (2.1).

Truncate A truncated distribution arises from restricting the support of a distribution.

$$\text{Truncated}_X(x \; ; a, b) = \frac{f(x)}{|F(a) - F(b)|}$$

The truncation of a continuous, univariate, unimodal distribution is also continuous, univariate and unimodal. Examples include the **Gompertz** distribution (a left-truncated Gumbel (8.5) distribution) and the **truncated normal distribution**.

Dual We create a dual distribution by interchanging the role of a variable and parameter in the probability density function.

$$Z(z \; ; x) = \frac{X(x \; ; z)}{\int dz\, X(x \; ; z)}$$

The integral (or sum, if z takes discrete values) in the denominator ensures that the dual distribution is normalized.

Tilt (exponential tilt, Esscher transform, exponential change of measure (ECM), twist) [169, 170]

$$\text{Tilted}_\theta\left(f(x)\right) = \frac{f(x)e^{\theta x}}{\int f(x)e^{\theta x}dx} = f(x)e^{\theta x - \kappa(\theta)}$$

Here $\kappa(\theta) = \ln \int f(x)e^{\theta x}dx$ is the cumulant generating function (p.158).

Generation

For an introduction to uniform random generation see Knuth [171], and for generating non-uniform variates from uniform random numbers see Devroye (1986) [42].

Fast, high quality algorithms are widely available for uniform random variables (e.g. the Mersenne Twister [172]), for the gamma distribution (e.g. the Marsaglia-Tsang fast gamma method [173]) and normal distributions (e.g. the ziggurat algorithm of Marsaglia and Tsang (2000) [174]). The exponential (§2), Laplace (§3) and power function (§5) distributions can be obtained from straightforward transformations of the uniform distribution.

The remaining simple distributions can be obtained from transforms of 1 or 2 gamma random variables [42] (See gamma distribution interrelations, (§7), p53), with the exception of the Pearson IV distribution, which can be sampled with a rejection method [42, 103].

F   Miscellaneous mathematics

Special functions

Gamma function [62]:

$$\Gamma(a) = \int_0^\infty t^{a-1}e^{-t}dt$$
$$= (a-1)!$$
$$= (a-1)\Gamma(a-1)$$

$$\Gamma(\tfrac{1}{2}) = \sqrt{\pi}$$
$$\Gamma(1) = 1$$
$$\Gamma(\tfrac{3}{2}) = \frac{\sqrt{\pi}}{2}$$
$$\Gamma(2) = 1$$

Incomplete gamma function [62]:

$$\Gamma(a,z) = \int_z^\infty t^{a-1}e^{-t}dt$$

$$\Gamma(a,0) = \Gamma(a)$$
$$\Gamma(1,z) = \exp(-x)$$
$$\Gamma(\tfrac{1}{2},z) = \sqrt{\pi}\,\mathrm{erfc}(\sqrt{z})$$

Regularized gamma function [62]:

$$Q(a;z) = \frac{\Gamma(a;z)}{\Gamma(a)}$$

$$Q(\tfrac{1}{2};z) = \mathrm{erfc}(\sqrt{z})$$
$$Q(1;z) = \exp(-z)$$
$$\tfrac{d}{dz}Q(a;z) = -\tfrac{1}{\Gamma(a)}z^{a-1}e^{-z}$$

Beta function [62]:

$$B(a, b) = \int_0^1 t^{a-1}(1 - t)^{b-1} dt$$
$$= \frac{\Gamma(a)\Gamma(b)}{\Gamma(a + b)}$$

$$B(a, b) = B(b, a)$$
$$B(1, b) = \tfrac{1}{b}$$
$$B(\tfrac{1}{2}, \tfrac{1}{2}) = \pi$$

When $a = b$ we have a *central beta function* [175].

Incomplete beta function [62]:

$$B(a, b; z) = \int_0^z t^{a-1}(1 - t)^{b-1} dt$$

$$\tfrac{d}{dz} B(a, b; z) = z^{a-1}(1 - z)^{b-1}$$
$$B(1, 1; z) = z$$

Regularized beta function [62]:

$$I(a, b; z) = \frac{B(a, b; z)}{B(a, b)}$$

$$I(a, b; 0) = 0$$
$$I(a, b; 1) = 1$$
$$I(a, b; z) = 1 - I(b, a; 1 - z)$$

Error function [62]:

$$\mathrm{erf}(z) = \frac{2}{\sqrt{\pi}} \int_0^z e^{-t^2} dt$$

Complimentary error function [62]:

$$\text{erfc}(z) = 1 - \text{erf}(z)$$
$$= \frac{2}{\sqrt{\pi}} \int_z^\infty e^{-t^2} \, dt.$$

Gudermannian function [62]:

$$\text{gd}(z) = \int_0^z \text{sech}(t) \, dt$$
$$= 2 \arctan(e^x) - \frac{\pi}{2}$$

A sinusoidal function.

Modified Bessel function of the first kind [62]:

$$I_\nu(z) = \left(\tfrac{1}{2}z\right)^\nu \sum_{k=0}^\infty \frac{\left(\tfrac{1}{4}z^2\right)^k}{k! \, \Gamma(\nu + k + 1)}$$

A monotonic, exponentially growing function.

Modified Bessel function of the second kind [62]:

$$K_\nu(z) = \frac{\pi}{2} \frac{I_{-\nu}(z) - I_\nu(z)}{\sin(\nu\pi)}$$

Another monotonic, exponentially growing function.

Arcsine function :

$$\arcsin(z) = \int_0^z \frac{1}{\sqrt{1 - x^2}} dx$$
$$\arcsin(\sin(z)) = z$$
$$\tfrac{d}{dz} \arcsin(z) = \frac{1}{\sqrt{1 - z^2}}$$

The functional inverse of the sin function.

Arctangent function:

$$\arctan(z) = \tfrac{1}{2}i\ln\frac{1-iz}{1+iz}$$

$$\arctan(z) = \int_0^z \frac{1}{1+x^2}dx$$

$$\arctan(\tan(z)) = z$$

$$\tfrac{d}{dz}\arctan(z) = \frac{1}{1+z^2}$$

$$\arctan(z) = -\arctan(-z)$$

The functional inverse of the tangent function.

Hyperbolic sine function:

$$\sinh(z) = \frac{e^{+x}-e^{-x}}{2}$$

Hyperbolic cosine function:

$$\cosh(z) = \frac{e^{+x}+e^{-x}}{2}$$

Hyperbolic secant function:

$$\mathrm{sech}(z) = \frac{2}{e^{+x}+e^{-x}} = \frac{1}{\cosh(z)}$$

Hyperbolic cosecant function:

$$\mathrm{csch}(z) = \frac{2}{e^{+x}-e^{-x}} = \frac{1}{\sinh(z)}$$

Hypergeometric function [62, 176]: All of the preceding functions can be expressed in terms of the hypergeometric function:

$$_pF_q(a_1, a_2, \ldots, a_p; b_1, b_2, \ldots, b_q; z) = \sum_{n=0}^{\infty} \frac{a_1^{\bar{n}}, \ldots, a_p^{\bar{n}}}{b_1^{\bar{n}}, \ldots, b_q^{\bar{n}}} \frac{z^n}{n!}$$

where $x^{\bar{n}}$ are rising factorial powers [62, 176]

$$x^{\bar{n}} = x(x+1)\cdots(x+n-1) = \frac{(x+n-1)!}{(x-1)!} \; .$$

The most common variant is $_2F_1(a, b; c; z)$, the Gauss hypergeometric function, which can also be defined using an integral formula due to Euler,

$$_2F_1(a, b; c; z) = \frac{1}{B(b, c-b)} \int_0^1 \frac{t^{b-1}(1-t)^{c-b-1}}{(1-zt)^a} dt \qquad |z| \leqslant 1 \, .$$

The variant $_1F_1(a; c; z)$ is called the confluent hypergeometric function, and $_0F_1(c; z)$ the confluent hypergeometric limit function.

Special cases include,

$$B(a, b; z) = \frac{z^a}{a} \, _2F_1(a, 1-b; a+1; z)$$

$$B(a, b) = \frac{1}{a} \, _2F_1(a, 1-b; a+1; 1)$$

$$\Gamma(a; z) = \Gamma(a) - \frac{z^a}{a} \, _1F_1(a; a+1; -z)$$

$$\mathrm{erfc}(z) = \frac{2z}{\sqrt{\pi}} \, _1F_1(\tfrac{1}{2}; \tfrac{3}{2}; -z^2)$$

$$\sinh(z) = z \, _0F_1(; \tfrac{3}{2}; \tfrac{z^2}{4})$$

$$\cosh(z) = \, _0F_1(; \tfrac{1}{2}; \tfrac{z^2}{4})$$

$$\arctan(z) = z \, _2F_1(\tfrac{1}{2}, 1; \tfrac{3}{2}; -z^2)$$

$$\arcsin(z) = z \, _2F_1(\tfrac{1}{2}, \tfrac{1}{2}; \tfrac{3}{2}; z^2)$$

$$I_\nu(z) = \frac{(\tfrac{1}{2}\nu)^\nu}{\Gamma(\nu+1)} \, _0F_1(; \nu+1; \tfrac{z^2}{4})$$

$$\tfrac{d}{dz} \, _2F_1(a, b; c; z) = \tfrac{ab}{c} \, _2F_1(a+1, b+1; c+1; z)$$

Sign function: The sign of the argument. For real arguments, the sign function is defined as

$$\mathrm{sgn}(x) = \begin{cases} -1 & \text{if } x < 0 \\ 0 & \text{if } x = 0 \, , \\ +1 & \text{if } x > 0 \end{cases}$$

and for complex arguments the sign function can be defined as

$$\operatorname{sgn}(z) = \begin{cases} \frac{z}{|z|} & \text{if } z \neq 0 \\ 0 & \text{if } z = 0 \end{cases}.$$

Polygamma function [62]: The $(n+1)$th logarithmic derivative of the gamma function. The first derivative is called the the **digamma function** (or psi function) $\psi(x) \equiv \psi_0(x)$, and the second the **trigamma function** $\psi_1(x)$.

$$\psi_n(x) = \frac{d^{n+1}}{dz^{n+1}} \ln \Gamma(x)$$
$$= \frac{d^n}{dz^n} \psi(x)$$

q-exponential and q-logarithmic functions [177, 178]: Two common and important limits are

$$\lim_{c \to 0} \frac{x^c - 1}{c} = \ln x$$

and

$$\lim_{c \to +\infty} \left(1 + \frac{x}{c}\right)^{ac} = e^{ax}.$$

It is sometimes useful to introduce 'q-deformed' exponential and logarithmic functions that extrapolate across these limits [177, 178].

$$\exp_q(x) = \begin{cases} \exp(x) & q = 1 \\ (1 + (1-q)x)^{\frac{1}{1-q}} & q \neq 1, \quad 1 + (1-q)x > 0 \\ 0 & q < 1, \quad 1 + (1-q)x \leqslant 0 \\ +\infty & q > 1, \quad 1 + (1-q)x \leqslant 0 \end{cases}$$

$$\ln_q(x) = \begin{cases} \frac{x^{1-q} - 1}{1-q} & q \neq 1 \\ \ln(x) & q = 1 \end{cases}$$

Note that these q-functions are unrelated to the q-exponential function defined in combinatorial mathematics.

Bibliography

[1] Gavin E. Crooks. Field guide to continuous probability distributions. `http://threeplusone.com/fieldguide` [Recursive citations mark neologisms and other innovations]. (pages 40, 40, 68, 68, 104, 104, 110, 127, 139, 140, 141, 142, 143, 155, 170, 195, 195, 195, 196, 196, 196, 196, 198, 199, 199, 199, 199, 199, 200, 200, 202, 203, 203, 205, and 205).

[2] Norman L. Johnson, Samuel Kotz, and Narayanaswamy Balakrishnan. *Continuous univariate distributions*, volume 1. Wiley, New York, 2nd edition (1994). (pages 4, 27, 33, 37, 37, 37, 37, 44, 47, 47, 48, 50, 72, 76, 77, 77, 77, 79, 79, 79, 88, 108, 109, 129, 133, 136, 146, 150, 154, and 199).

[3] Norman L. Johnson, Samuel Kotz, and Narayanaswamy Balakrishnan. *Continuous univariate distributions*, volume 2. Wiley, New York, 2nd edition (1995). (pages 4, 36, 54, 55, 56, 57, 57, 59, 63, 63, 84, 85, 94, 97, 98, 108, 110, 111, 111, 111, 125, 142, 143, 146, 148, 148, 148, 198, and 198).

[4] Karl Pearson. Contributions to the mathematical theory of evolution. *Philos. Trans. R. Soc. A*, 54:329–333 (1893). doi:10.1098/rspl.1893.0079. (pages 47 and 132).

[5] Karl Pearson. Contributions to the mathematical theory of evolution. II. Skew variation in homogeneous material. *Philos. Trans. R. Soc. A*, 186:343–414 (1895). doi:10.1098/rsta.1895.0010. (pages 47, 89, 93, 114, 129, 132, 132, 132, and 132).

[6] Karl Pearson. Mathematical contributions to the theory of evolution. X. Supplement to a memoir on skew variation. *Philos. Trans. R. Soc. A*, 197:443–459 (1901). doi:10.1098/rsta.1901.0023. (pages 79, 97, 97, 129, 132, and 132).

[7] Karl Pearson. Mathematical contributions to the theory of evolution. XIX. Second supplement to a memoir on skew variation. *Philos. Trans. R. Soc. A*, 216:429–457 (1916). doi:10.1098/rsta.1916.0009. (pages 27, 36, 37, 37, 37, 60, 91, 91, 129, 132, 132, 132, 132, 132, 132, and 203).

[8] Lawrence M. Leemis and J. T. McQueston. Univariate distribution relationships. *Amer. Statistician*, 62:45–53 (2008). doi:10.1198/000313008X270448. (pages 4, 117, and 202).

[9] Lawrence M. Leemis, Daniel J. Luckett, Austin G. Powell, and Peter E. Vermeer. Univariate probability distributions. *J. Stat. Edu.*, 20(3) (2012). `http://www.math.wm.edu/~leemis/chart/UDR/UDR.html`. (page 4).

[10] Gavin E. Crooks. The Amoroso distribution. arXiv:1005.3274v2. (page 5).

[11] T. Kondo. A theory of sampling distribution of standard deviations. *Biometrika*, 22:36–64 (1930). doi:10.1093/biomet/22.1-2.36. (page 27).

[12] Pierre Simon Laplace. Memoir on the probability of the causes of events (1774). doi:10.1214/ss/1177013621. (page 30).

[13] Stephen M. Stigler. Laplace's 1774 memoir on inverse probability. *Statist. Sci.*, 1(3):359–363 (1986). (page 30).

[14] Samuel Kotz, T. J. Kozubowski, and K. Podgórski. *The Laplace distribution and generalizations: A revisit with applications to communications, economics, engineering, and finance.* Birkhäuser, Boston (2001). (page 30).

[15] Abraham de Moivre. *The doctrine of chances.* Woodfall, London, 2nd edition (1738). (page 33).

[16] George E. P. Box and Mervin E. Muller. A note of the generation of random normal deviates. *Ann. Math. Statist.*, 29:610–611 (1958). doi:10.1214/aoms/1177706645. (page 35).

[17] M. Meniconi and D. M. Barry. The power function distribution: A useful and simple distribution to assess electrical component reliability. *Microelectron. Reliab.*, 36(9):1207–1212 (1996). doi:10.1016/0026-2714(95)00053-4. (page 36).

[18] Vilfredo Pareto. *Cours d'économie politique.* Librairie Droz, Geneva, Nouvelle édition par G.-H. Bousquet et G. Busino edition (1964). (page 37).

[19] K. S. Lomax. Business failures: Another example of the analysis of failure data. *J. Amer. Statist. Assoc.*, 49:847–852 (1954). doi:10.2307/2281544. (page 39).

[20] M. C. Jones. Families of distributions arising from distributions of order statistics. *Test*, 13(1):1–43 (2004). doi:10.1007/BF02602999. (pages 40, 113, and 201).

[21] P. C. Consul and G. C. Jain. On the log-gamma distribution and its properties. *Statistical Papers*, 12(2):100–106 (1971). doi:10.1007/BF02922944. (pages 42, 67, and 68).

[22] Francis Galton. The geometric mean, in vital and social statistics. *Proc. R. Soc. Lond.*, 29:365–267 (1879). doi:10.1098/rspl.1879.0060. (page 44).

[23] Donald McAlister. The law of the geometric mean. *Proc. R. Soc. Lond.*, 29:367–376 (1879). doi:10.1098/rspl.1879.0061. (page 44).

[24] R. Gibrat. *Les inégalités économiques.* Librairie du Recueil Sirey, Paris (1931). (page 44).

[25] A. K. Erlang. The theory of probabilities and telephone conversations. *Nyt Tidsskrift for Matematik B*, 20:33–39 (1909). (page 47).

[26] Ronald A. Fisher. On a distribution yielding the error functions of several well known statistics. In *Proceedings of the International Congress of Mathematics, Toronto*, volume 2, pages 805–813 (1924). (page 48).

[27] Peter M. Lee. *Bayesian Statistics: An Introduction*. Wiley, New York, 4th edition (2012). (pages 49, 55, 81, and 82).

[28] C. E. Porter and R. G. Thomas. Fluctuations of nuclear reaction widths. *Phys. Rev.*, 104:483–491 (1956). doi:10.1103/PhysRev.104.483. (page 50 and 50).

[29] M. S. Bartlett and M. G. Kendall. The statistical analysis of variance-heterogeneity and the logarithmic transformation. *J. Roy. Statist. Soc. Suppl.*, 8(1):128–138 (1946). http://www.jstor.org/stable/2983618. (page 54).

[30] Ross L. Prentice. A log gamma model and its maximum likelihood estimation. *Biometrika*, 61:539–544 (1974). doi:10.1093/biomet/61.3.539. (page 54).

[31] Eduardo Gutiérrez González, José A. Villaseñor Alva, Olga V. Panteleeva, and Humberto Vaquera Huerta. On testing the log-gamma distribution hypothesis by bootstrap. *Comp. Stat.*, 28(6):2761–2776 (2013). doi:10.1007/s00180-013-0427-4. (pages 54 and 72).

[32] Emil J. Gumbel. *Statistics of extremes*. Columbia Univ. Press, New York (1958). (pages 55, 57, 57, 57, 84, 84, and 86).

[33] Ronald A. Fisher and Leonard H. C. Tippett. Limiting forms of the frequency distribution of the largest or smallest member of a sample. *Proc. Camb. Phil. Soc.*, 24:180–190 (1928). doi:10.1017/S0305004100015681. (pages 57, 84, 148, 148, and 148).

[34] S. T. Bramwell, P. C. W. Holdsworth, and J.-F. Pinton. Universality of rare fluctuations in turbulence and citical phenomena. *Nature*, 396:552–554 (1998). (page 58).

[35] S. T. Bramwell, K. Christensen, J.-Y. Fortin, P. C. W. Holdsworth, H. J. Jensen, S. Lise, J. M. López, M. Nicodemi, J.-F. Pinton, and M. Sellitto. Universal fluctuations in correlated systems. *Phys. Rev. Lett.*, 84:3744–3747 (2000). (page 58).

[36] José E. Moyal. XXX. Theory of ionization fluctuations. *The London, Edinburgh, and Dublin Philosophical Magazine and Journal of Science*, 46(374):263–280 (1955). doi:10.1080/14786440308521076. (page 58 and 58).

[37] Student. The probable error of mean. *Biometrika*, 6:1–25 (1908). (pages 60, 62, and 63).

[38] Ronald A. Fisher. Applications of "Student's" distribution. *Metron*, 5:90–104 (1925). (pages 60 and 63).

[39] J. A. Hanley, M. Julien, and E. E. M. Moodie. Student's z, t, and s: What if Gosset had R? *Amer. Statistician*, 62:64–69 (2008). doi:10.1198/000313008X269602. (pages 60 and 62).

[40] Sandy L. Zabell. On Student's 1908 article "The probable error of a mean". *J. Amer. Statist. Assoc.*, 103:1–7 (2008). doi:10.1198/016214508000000030. (page 60).

[41] M. C. Jones. Student's simplest distribution. *Statistician*, 51:41–49 (2002). doi:10.1111/1467-9884.00297. (page 61).

[42] Luc Devroye. *Non-uniform random variate generation*. Springer-Verlag, New York (1986). http://www.nrbook.com/devroye/. (pages 62, 62, 65, 65, 140, 168, 174, 174, 174, and 199).

[43] Siméon Denis Poisson. Sur la probabilité des résultats moyens des observation. *Connaissance des Temps pour l'an*, pages 273–302 (1827). (page 63).

[44] A. L. Cauchy. Sur les résultats moyens d'observations de même nature, et sur les résultats les plus probables. *Comptes Rendus de l'Académie des Sciences*, 37:198–206 (1853). (page 63).

[45] Gregory Breit and Eugene Wigner. Capture of slow neutrons. *Phys. Rev.*, 49(7):519–531 (1936). doi:10.1103/PhysRev.49.519. (page 64).

[46] A. C. Olshen. Transformations of the Pearson type III distribution. *Ann. Math. Statist.*, 9:176–200 (1938). http://www.jstor.org/stable/2957731. (page 67).

[47] A. Grassia. On a family of distributions with argument between 0 and 1 obtained by transformation of the gamma and derived compound distributions. *Aust. J. Statist.*, 19(2):108–114 (1977). doi:10.1111/j.1467-842X.1977.tb01277.x. (pages 67 and 71).

[48] A. K. Gupta and S. Nadarajah, editors. *Handbook of beta distribution and its applications*. Marcel Dekker, New York (2004). (page 67).

[49] Melvin D. Springer. *The algebra of random variables*. John Wiley (1979). (pages 67, 151, and 168).

[50] Luigi Amoroso. Richerche intorno alla curve die redditi. *Ann. Mat. Pura Appl.*, 21:123–159 (1925). (page 72 and 72).

[51] James B. McDonald. Some generalized functions for the size distribution of income. *Econometrica*, 52(3):647–663 (1984). (pages 72, 86, 117, 119, 121, 122, 122, 124, 128, 195, 195, and 196).

[52] E. W. Stacy. A generalization of the gamma distribution. *Ann. Math. Statist.*, 33(3):1187–1192 (1962). (page 73).

[53] Ali Dadpay, Ehsan S. Soofi, and Refik Soyer. Information measures for generalized gamma family. *J. Econometrics*, 138:568–585 (2007). doi:10.1016/j.jeconom.2006.05.010. (pages 73 and 87).

[54] Viorel Gh. Vodă. New models in durability tool-testing: pseudo-Weibull distribution. *Kybernetika*, 25(3):209–215 (1989). (page 73).

[55] Wenhao Gui. Statistical inferences and applications of the half exponential power distribution. *J. Quality and Reliability Eng.*, 2013:219473 (2013). doi:10.1155/2013/219473. (page 75).

[56] Evan Hohlfeld and Phillip L. Geissler. Dominance of extreme statistics in a prototype many-body brownian ratchet. *J. Chem. Phys.*, 141:161101 (2014). doi:10.1063/1.4899052. (page 75 and 75).

[57] Minoru Nakagami. The m-distribution – A general formula of intensity distribution of rapid fading. In W. C. Hoffman, editor, *Statistical methods in radio wave propagation: Proceedings of a symposium held at the University of California, Los Angeles, June 18-20, 1958*, pages 3–36. Pergamon, New York (1960). doi:10.1016/B978-0-08-009306-2.50005-4. (pages 75, 75, and 201).

[58] K. S. Miller. *Multidimensional Gaussian distributions*. Wiley, New York (1964). (page 77).

[59] John W. Strutt (Lord Rayleigh). On the resultant of a large number of vibrations of the same pitch and of arbitrary phase. *Phil. Mag.*, 10:73–78 (1880). doi:10.1080/14786448008626893. (page 77).

[60] C. G. Justus, W. R. Hargraves, A. Mikhail, and D. Graberet. Methods for estimating wind speed frequency distributions. *J. Appl. Meteorology*, 17(3):350–353 (1978). (page 78).

[61] James C. Maxwell. Illustrations of the dynamical theory of gases. Part 1. On the motion and collision of perfectly elastic spheres. *Phil. Mag.*, 19:19–32 (1860). (page 78).

[62] Milton Abramowitz and Irene A. Stegun. *Handbook of mathematical functions with formulas, graphs, and mathematical tables*. Dover, New York (1965). (pages 78, 175, 175, 175, 176, 176, 176, 176, 177, 177, 177, 177, 178, 179, and 180).

[63] Edwin B. Wilson and Margaret M. Hilferty. The distribution of chi-square. *Proc. Natl. Acad. Sci. U.S.A.*, 17:684–688 (1931). (page 79 and 79).

[64] D. M. Hawkins and R. A. J. Wixley. A note on the transformation of chi-squared variables to normality. *Amer. Statistician*, 40:296–298 (1986). doi:10.2307/2684608. (page 79).

[65] Andrew Gelman, John B. Carlin, Hal S. Stern, and Donald B. Rubin. *Bayesian data analysis*. Chapman and Hall, New York, 2nd edition (2004). (pages 79, 81, 81, 81, 96, 153, and 203).

[66] C. Kleiber and Samuel Kotz. *Statistical size distributions in economics and actuarial sciences*. Wiley, New York (2003). (pages 79, 99, 123, and 125).

[67] W. Feller. *An introduction to probability theory and its applications*, volume 2. Wiley, New York, 2nd edition (1971). (pages 80, 80, and 122).

[68] M. Evans, N. Hastings, and J. B. Peacock. *Statistical distributions*. Wiley, New York, 3rd edition (2000). (pages 82 and 151).

[69] Viorel Gh. Vodă. On the inverse Rayleigh random variable. *Rep. Statist. Appl. Res., JUSE*, 19:13–21 (1972). (page 82).

[70] Mohammad Shakil, B. M. Golam Kibria, and Jai Narain Singh. A new family of distributions based on the generalized Pearson differential equation with some applications. *Aust. J. Statist.*, 39(3):259–278 (2010). (pages 82, 82, 139, 198, 199, 199, 199, and 200).

[71] Francisco Louzada, Pedro Luiz Ramos, and Diego Nascimento. The inverse Nakagami-m distribution: A novel approach in reliability. *IEEE Trans. Reliability*, 67(3):1030–1042 (2018). doi:10.1109/TR.2018.2829721. (pages 83 and 200).

[72] Nikolaĭ V. Smirnov. Limit distributions for the terms of a variational series. *Trudy Mat. Inst. Steklov.*, 25:3–60 (1949). (pages 83, 85, and 86).

[73] Ole E. Barndorff-Nielsen. On the limit behaviour of extreme order statistics. *Ann. Math. Statist.*, 34:992–1002 (1963). (pages 83, 85, and 86).

[74] Richard von Mises. La distribution de la plus grande de n valeurs. *Rev. Math. Union Interbalcanique*, 1:141–160 (1936). (page 84 and 84).

[75] Daniel McFadden. Modeling the choice of residential location. *Transportation Research Record*, 673:72–77 (1978). (page 84).

[76] W. Weibull. A statistical distribution function of wide applicability. *J. Appl. Mech.*, 18:293–297 (1951). (page 85).

[77] M. Fréchet. Sur la loi de probabilité de l'écart maximum. *Ann. Soc. Polon. Math.*, 6:93–116 (1927). (page 86).

[78] J. F. Lawless. *Statistical models and methods for lifetime data*. Wiley, New York (1982). (page 88).

[79] Charles E. Clark. The PERT model for the distribution of an activity. *Operations Research*, 10:405–406 (1962). doi:10.1287/opre.10.3.405. (page 90).

[80] D. Vose. *Risk analysis - A quantitative guide*. John Wiley & Sons, New York, 2nd edition (2000). (pages 90 and 91).

[81] Robert M. Norton. On properties of the arc-sine law. *Sankhyā*, A37:306–308 (1975). http://www.jstor.org/stable/25049987. (pages 93, 94, 94, and 94).

[82] W. Feller. *An introduction to probability theory and its applications*, volume 1. Wiley, New York, 3rd edition (1968). (page 94).

[83] Eugene P. Wigner. Characteristic vectors of bordered matrices with infinite dimensions. *Ann. Math.*, 62:548–564 (1955). (page 94).

[84] V. A. Epanechnikov. Non-parametric estimation of a multivariate probability density. *Theory Probab. Appl.*, 14:153–158 (1969). doi:10.1137/1114019. (page 94).

[85] R. Durbin, S. R. Eddy, A. Krogh, and G. Mitchison. *Biological sequence analysis*. Cambridge University Press, Cambridge (1998). (page 96).

[86] George W. Snedecor. *Calculation and interpretation of analysis of variance and covariance*. Collegiate Press, Ames, Iowa (1934). (page 98).

[87] Leo A. Aroian. A study of R. A. Fisher's z distribution and the related F distribution. *Ann. Math. Statist.*, 12:429–448 (1941). http://www.jstor.org/stable/2235955. (page 98).

[88] Satya D. Dubey. Compound gamma, beta and F distributions. *Metrika*, 16(1):27–31 (1970). doi:10.1007/BF02613934. (pages 101 and 197).

[89] J. C. Ahuja and Stanley W. Nash. The generalized Gompertz-Verhulst family of distributions. *Sankhyā*, 29:141–156 (1967). http://www.jstor.org/stable/25049460. (pages 102, 102, 198, 198, 198, and 199).

[90] Saralees Nadarajah and Samuel Kotz. The beta exponential distribution. *Reliability Eng. Sys. Safety*, 91:689–697 (2006). doi:10.1016/j.ress.2005.05.008. (pages 102, 104, 106, 106, 106, 106, 106, 106, and 106).

[91] Srividya Iyer-Biswas, Gavin E. Crooks, Norbert F. Scherer, and Aaron R. Dinner. Universality in stochastic exponential growth. *Phys. Rev. Lett.*, 113:028101 (2014). doi:10.1103/PhysRevLett.113.028101. (page 102).

[92] Pierre François Verhulst. Deuxième mémoire sur la loi d'accroissement de la population. *Mém. de l'Academie Royale des Sci., des Lettres et des Beaux-Arts de Belgique*, 20(1–32) (1847). (page 102).

[93] Rameshwar D. Gupta and Debasis Kundu. Exponentiated exponential family: An alternative to gamma and Weibull distributions. *Biometrical J.*, 1:117–130 (2001). (page 102).

[94] Ross L. Prentice. A generalization of probit and logit methods for dose response curves. *Biometrics*, 32(4):761–768 (1976). (page 108).

[95] James B. McDonald. Parametric models for partially adaptive estimation with skewed and leptokurtic residuals. *Econ. Lett.*, 37:273–278 (1991). (pages 108 and 144).

[96] Richard Morton, Peter C. Annis, and Helen A. Dowsett. Exposure time in low oxygen for high mortality of *sitophilus oryzae* adults: An application of generalized logit models. *J. Agri. Biol. Env. Stat.*, 5:360–371 (2000). (pages 108 and 203).

[97] Irving W. Burr. Cumulative frequency functions. *Ann. Math. Statist.*, 13:215–232 (1942). doi:10.1214/aoms/1177731607. (pages 108, 123, and 125).

[98] P.-F Verhulst. Recherches mathématiques sur la loi d'accroissement de la populationmatiques sur la loi d'accroissement de la population. *Nouv. mém. de l'Academie Royale des Sci. et Belles-Lettres de Bruxelles*, 18:1–41 (1845). (page 111).

[99] Narayanaswamy Balakrishnan. *Handbook of the logistic distribution*. CRC Press, New York (1991). (page 111).

[100] Wilfred F. Perks. On some experiments in the graduation of mortality statistics. *J. Inst. Actuar.*, 63:12–57 (1932). http://www.jstor.org/stable/41137425. (pages 111 and 140).

[101] J. Talacko. Perks' distributions and their role in the theory of Wiener's stochastic variables. *Trabajos de Estadistica*, 7:159–174 (1956). doi:10.1007/BF03003994. (pages 111 and 140).

[102] Allen Birnbaum and Jack Dudman. Logistic order statistics. *Ann. Math. Statist.*, 34(658–663) (1963). (page 113).

[103] J. Heinrich. A guide to the Pearson type IV distribution. CDF/memo/statistic/public/6820. (pages 114 and 174).

[104] Poondi Kumaraswamy. A generalized probability density function for double-bounded random processes. *J. Hydrology*, 46:79–88 (1980). doi:10.1016/0022-1694(80)90036-0. (page 117).

[105] M. C. Jones. Kumaraswamy's distribution: A beta-type distribution with some tractability advantages. *Statistical Methodology*, 6:70–81 (2009). doi:10.1016/j.stamet.2008.04.001. (page 117).

[106] S. Tahmasebi and J. Behboodian. Shannon entropy for the Feller-Pareto (FP) family and order statistics of FP subfamilies. *Appl. Math. Sci.*, 4:495–504 (2010). (page 122).

[107] S. A. Klugman, H. H. Panjer, and G. E. Willmot. *Loss models: From data to decisions*. Wiley, New York, 3rd edition (2004). (pages 122 and 125).

[108] Pandu R. Tadikamalla. A look at the Burr and related distributions. *Int. Stat. Rev.*, 48(3):337–344 (1980). doi:10.2307/1402945. (pages 123 and 125).

[109] Camilo Dagum. A new model of personal income distribution: Specification and estimation. *Economie Appliquée*, 30:413–437 (1977). (page 125).

[110] B. K. Shah and P. H. Dave. A note on log-logistic distribution. *J. Math. Sci. Univ. Baroda (Sci. Number)*, 12:21–22 (1963). (page 125).

[111] Andrew Gelman. Prior distributions for variance parameters in hierarchical models. *Bayesian Analysis*, 3:515–533 (2006). (page 126, 126, and 126).

[112] J. K. Ord. *Families of frequency distributions*. Griffin, London (1972). (page 129 and 129).

[113] Hans van Leeuwen and Hans Maassen. A q deformation of the Gauss distribution. *J. Math. Phys.*, 36(9):4743–4756 (1995). doi:10.1063/1.530917. (page 131).

[114] L. K. Roy. An extension of the Pearson system of frequency curves. *Trabajos de estadistica y de investigacion operativa*, 22(1–2):113–123 (1971). (page 133).

[115] Abraham Wald. On cumulative sums of random variables. *Ann. Math. Statist.*, 15(3):283–296 (1944). doi:10.1214/aoms/1177731235. (page 133).

[116] Maurice C. K. Tweedie. Inverse statistical variates. *Nature*, 155:453 (1945). doi:10.1038/155453a0. (page 133).

[117] J. Leroy Folks and Raj S. Chhikara. The inverse Gaussian distribution and its statistical application - A review. *J. Roy. Statist. Soc. B*, 40:263–289 (1978). http://www.jstor.org/stable/2984691. (page 133).

[118] Raj S. Chhikara and J. Leroy Folks. *The inverse Gaussian distribution: Theory, methodology, and applications*. Marcel Dekker, New York (1988). (pages 133 and 135).

[119] Étienne Halphen. Sur un nouveau type de courbe de fréquence. *Comptes Rendus de l'Académie des Sciences*, 213:633–635 (1941). Published under the name of "Dugué". (pages 136, 136, and 137).

[120] Luc Perreault, Bernard Bobée, and Peter F. Rasmussen. Halphen distribution system. I: Mathematical and statistical properties. *J. Hydrol. Eng.*, 4:189–199 (1999). doi:10.1061/(ASCE)1084-0699(1999)4:3(189). (pages 136, 137, 137, and 137).

[121] Étienne Halphen. Les fonctions factorielles. *Publications de l'Institut de Statistique de l'Université de Paris*, 4(1):21–39 (1955). (page 137).

[122] G. Morlat. Les lois de probabilités de Halphen. *Revue de Statistique Appliquée*, 3:21–46 (1956). (page 137).

[123] I. J. Good. The population frequencies of species and the estimation of population parameters. *Biometrika*, 40:237–264 (1953). doi:10.1093/biomet/40.3-4.237. (page 137).

[124] Herbert S. Sichel. Statistical valuation of diamondiferous deposits. *J. South. Afr. Inst. Min. Metall.*, pages 235–243 (1973). (page 137).

[125] Ole E. Barndorff-Nielsen. Exponentially decreasing distributions for the logarithm of particle size. *Proc. Roy. Soc. London*, 353:401–419 (1977). doi:10.1098/rspa.1977.0041. (page 137).

[126] David L. Libby and Melvin R. Novick. Multivariate generalized beta-distributions with applications to utility assessment. *J. Edu. Stat.*, 7:271–294 (1982). (page 138).

[127] James B. McDonald and Yexiao J. Xu. A generalization of the beta distribution with applications. *J. Econometrics*, 66:133–152 (1995). (pages 138, 197, 197, 198, and 201).

[128] José María Sarabia and Enrique Castillo. *Advances in Distribution Theory, Order Statistics, and Inference*, chapter Bivariate distributions based on the generalized three-parameter beta distribution, pages 85–110. Birkhäuser, Boston, MA (2006). (page 138, 138, and 138).

[129] Saralees Nadarajah and Samuel Kotz. Multitude of beta distributions with applications. *Statistics*, 41:153–179 (2007). doi:10.1080/02331880701223522. (pages 138 and 139).

[130] Carmen Armero and Maria J. Bayarri. Prior assessments for prediction in queues. *The Statistician*, 43(1):139–153 (1994). http://www.jstor.org/stable/2348939. (page 138).

[131] Michael B. Gordy. Computationally convenient distributional assumptions for common-value auctions. *Computational Economics*, 12:61–78 (1988). (page 139 and 139).

[132] Saralees Nadarajah and Samuel Kotz. An F_1 beta distribution with bathtub failure rate function. *Amer. J. Math. Manag. Sci.*, 26(1-2):113–131 (2006). doi:10.1080/01966324.2006.10737663. (pages 139, 141, and 195).

[133] D. G. Champernowne. The graduation of income distributions. *Econometrica*, 20(4):591–615 (1952). http://www.jstor.org/stable/1907644. (page 140).

[134] P. R. Rider. Generalized Cauchy distributions. *Ann. Inst. Statist. Math.*, 9(1):215–223 (1958). doi:10.1007/BF02892507. (pages 141, 144, and 204).

[135] R. G. Laha. An example of a nonnormal distribution where the quotient follows the Cauchy law. *Proc. Natl. Acad. Sci. U.S.A.*, 44(2):222–223 (1958). doi:10.1214/aoms/1177706102. (page 141).

[136] Ileana Popescu and Monica Dumitrescu. Laha distribution: Computer generation and applications to life time modelling. *J. Univ. Comp. Sci.*, 5:471–481 (1999). doi:10.3217/jucs-005-08-0471. (page 141 and 141).

[137] Grace E. Bates. Joint distributions of time intervals for the occurrence of succesive accidents in a generalized Polya urn scheme. *Ann. Math. Statist.*, 26(4):705–720 (1955). (page 142).

[138] Shelemyahu Zacks. Estimating the shift to wear-out systems having exponential-Weibull life distributions. *Operations Research*, 32(1):741–749 (1984). doi:10.1287/opre.32.3.741. (page 143).

[139] Govind S. Mudholkar, Deo Kumar Srivastava, and Marshall Freimer. The exponentiated Weibull family: A reanalysis of the Bus-Motor-Failure data. *Technometrics*, 37(4):436–445 (1995). (page 143).

[140] Z. W. Birnbaum and S. C. Saunders. A new family of life distributions. *J. Appl. Prob.*, 6(2):319–327 (1969). doi:10.2307/3212003. (page 143).

[141] George E. P. Box and George C. Tiao. A further look at robustness via Bayes's theorem. *Biometrika*, 49:419–432 (1962). doi:10.2307/2333976. (page 143).

[142] Saralees Nadarajah. A generalized normal distribution. *J. Appl. Stat.*, 32(7):685–694 (2005). doi:10.1080/02664760500079464. (page 143).

[143] Henry John Malik. Exact distribution of the product of independent generalized gamma variables with the same shape parameter. *Ann. Stat.*, 39:1751–1752 (1968). (pages 144, 145, and 146).

[144] James H. Miller and John B. Thomas. Detectors for discrete-time signals in non-Gaussian noise. *IEEE Trans. Inf. Theory*, 18(2):241–250 (1972). doi:10.1109/TIT.1972.1054787. (page 144).

[145] James B. McDonald and Whitney K. Newey. Partially adaptive estimation of regression models via the generalized t distribution. *Econometric Theory*, 4:428–457 (1988). doi:10.1017/S0266466600013384. http://www.jstor.org/stable/3532334. (page 144).

[146] Saralees Nadarajah and K. Zografos. Formulas for Rényi information and related measures for univariate distributions. *Information Sciences*, 155:119–138 (2003). doi:10.1016/S0020-0255(03)00156-7. (pages 144 and 196).

[147] Tuncer C. Aysal and Kenneth E. Barner. Meridian filtering for robust signal processing. *IEEE Trans. Signal. Process.*, 55(8):3949–3962 (2007). doi:10.1109/TSP.2007.894383. (pages 144, 147, and 147).

[148] J. Holtsmark. Über die Verbreiterung von Spektrallinien. *Ann. Phys.*, 363(7):577–630 (1919). doi:10.1002/andp.19193630702. (page 145).

[149] Timothy M. Garoni and Norman E. Frankel. Lévy flights: Exact results and asymptotics beyond all orders. *J. Math. Phys.*, 43(5):2670–2689 (2002). doi:10.1063/1.1467095. (page 145).

[150] E. Jakeman and P. N. Pusey. Significance of k-distributions in scattering experiments. *Phys. Rev. Lett.*, 40:546–550 (1978). doi:10.1103/PhysRevLett.40.546. (pages 145 and 146).

[151] Nicholas J. Redding. Estimating the parameters of the K distribution in the intensity domain (1999). Report DSTO-TR-0839, DSTO Electronics and Surveillance Laboratory, South Australia. (pages 145, 146, and 146).

[152] Christopher S. Withers and Saralees Nadarajah. On the product of gamma random variables. *Quality & Quantity*, 47(1):545–552 (2013). doi:10.1007/s11135-011-9474-5. (pages 145 and 146).

[153] Purushottam D. Dixit. A maximum entropy thermodynamics of small systems. *J. Chem. Phys.*, 138(18):184111 (2013). doi:http://dx.doi.org/10.1063/1.4804549. (page 146).

[154] J. O. Irwin. On the frequency distribution of the means of samples from a population having any law of frequency with finite moments, with special reference to Pearson's type II. *Biometrika*, 19:225–239 (1927). doi:10.2307/2331960. (page 146).

[155] Philip Hall. The distribution of means for samples of size n drawn from a population in which the variate takes values between 0 and 1, all such values being equally probable. *Biometrika*, 19:240–245 (1927). doi:10.2307/2331961. (page 146).

[156] Norman L. Johnson. Systems of frequency curves generated by methods of translation. *Biometrika*, 36:149–176 (1949). doi:10.2307/2332539. (page 146).

[157] Lev D. Landau. On the energy loss of fast particles by ionization. *J. Phys. (USSR)*, 8:201–205 (1944). (page 147 and 147).

[158] Albert W. Marshall and Ingram Olkin. *Life distributions. Structure of nonparametric, semiparametric and parametric families.* Springer (2007). (pages 147 and 205).

[159] G. K. Wertheim, M. A. Butler, K. W. West, and D. N. E. Buchanan. Determination of the Gaussian and Lorentzian content of experimental line shapes. *Rev. Sci. Instrum.*, 45:1369–1371 (1974). doi:10.1063/1.1686503. (page 149).

[160] S. O. Rice. Mathematical analysis of random noise. Part III. *Bell Syst. Tech. J.*, 24:46–156 (1945). doi:10.1002/j.1538-7305.1945.tb00453.x. (page 149).

[161] Kushal K. Talukdar and William D. Lawing. Estimation of the parameters of the Rice distribution. *J. Acoust. Soc. Am.*, 89(3):1193–1197 (1991). doi:10.1121/1.400532. (page 149).

[162] W. H. Rogers and J. W. Tukey. Understanding some long-tailed symmetrical distributions. *Statistica Neerlandica*, 26(3):211–226 (1972). doi:10.1111/j.1467-9574.1972.tb00191.x. (page 150).

[163] John P. Nolan. *Stable Distributions - Models for Heavy Tailed Data*. Birkhäuser, Boston (2015). (page 150).

[164] Makoto Yamazato. Unimodality of infinitely divisible distribution functions of class L. *Ann. Prob.*, 6(4):523–531 (1978). http://www.jstor.org/stable/2243119. (page 150).

[165] Hirofumi Suzuki. A statistical model for urban radio propagation. *IEEE Trans. Comm.*, 25(7):673–680 (1977). (page 151).

[166] B. H. Armstrong. Spectrum line profiles: The Voigt function. *J. Quant. Spectrosc. Radiat. Transfer*, 7:61–88 (1967). doi:10.1016/0022-4073(67)90057-X. (page 152 and 152).

[167] Herbert A. David and Haikady N. Nagaraja. *Order Statistics*. Wiley, 3rd edition (2005). doi:10.1002/0471722162. (page 161).

[168] N. Eugene, C. Lee, and F. Famoye. Beta-normal distributions and its applications. *Commun. Statist.-Theory Meth.*, 31(4):497–512 (2002). (page 161).

[169] Fredrik Esscher. On the probability function in the collective theory of risk. *Scandinavian Actuarial Journal*, 1932(3):175–195 (1932). doi:10.1080/03461238.1932.10405883. (page 174).

[170] D. Siegmund. Importance sampling in the Monte Carlo study of sequential tests. *Ann. Statist.*, 4(4):673–684 (1976). doi:10.1214/aos/1176343541. (page 174).

[171] Donald E. Knuth. *Art of computer programming, volume 2: Seminumerical algorithms*. Addison-Wesley, New York, 3rd edition (1997). (page 174).

[172] M. Matsumoto and T. Nishimura. Mersenne Twister: A 623-dimensionally equidistributed uniform pseudorandom number generator. *ACM Trans. Model. Comput. Simul.*, 8(1):3–30 (1998). doi:10.1145/272991.272995. (page 174).

[173] George Marsaglia and Wai Wan Tsang. A simple method for generating gamma variables. *ACM Trans. Math. Soft.*, 26(3):363–372 (2001). doi:10.1145/358407.358414. (page 174).

[174] George Marsaglia and Wai Wan Tsang. The ziggurat method for generating random variables. *J. Stat. Soft.*, 5(8):1–7 (2000). (page 174).

[175] J. M. Borwein and I. J. Zucker. Elliptic integral evaluation of the gamma function at rational values of small denominators. *IMA J. Numerical Analysis*, 12:519–526 (1992). (page 176).

[176] Ronald L. Graham, Donald E. Knuth, and Oren Patasknik. *Concrete mathematics: A foundation for computer science*. Addison-Wesley, 2nd edition (1994). (pages 178 and 179).

[177] Constantino Tsallis. What are the numbers that experiments provide? *Quimica Nova*, 17:468–471 (1994). (page 180 and 180).

[178] Takuya Yamano. Some properties of q-logarithm and q-exponential functions in Tsallis statistics. *Physica A*, 305(3–4):486–496 (2002). doi:10.1016/S0378-4371(01)00567-2. (page 180 and 180).

[179] John Aitchison and James A. C. Brown. *The lognormal distribution with special references to its uses in economics*. Cambridge University Press, Cambridge (1966). (page 195).

[180] Stephen M. Stigler. Cauchy and the Witch of Agnesi: An historical note on the Cauchy distribution. *Biometrika*, 61(2):375–380 (1974). http://www.jstor.org/stable/2334368. (pages 196 and 206).

[181] A. J. Coale and D. R. McNeil. The distribution by age of the frequency of first marriage in female cohort. *J. Amer. Statist. Assoc.*, 67:743–749 (1972). doi:10.2307/2284631. (page 196).

[182] Ganapati P. Patil, M. T. Boswell, and M. V. Ratnaparkhi. *Dictionary and classified bibliography of statistical distributions in scientific work: Continuous univariate models*. International co-operative publishing house (1984). (page 198).

[183] R. D. Gupta and D. Kundu. Generalized exponential distribution: Existing results and some recent developments. *J. Statist. Plann. Inference*, 137:3537–3547 (2007). doi:10.1016/j.jspi.2007.03.030. (pages 198 and 199).

[184] Barry C. Arnold. *Pareto distributions*. International co-operative publishing house (1983). (page 198).

[185] Carlos A. Coelho and João T. Mexia. On the distribution of the product and ratio of independent generalized gamma-ratio random variables. *Sankhyā*, 69(2):221–255 (2007). http://www.jstor.org/stable/25664553. (page 198).

[186] M. M. Hall, H. Rubin, and P. G. Winchell. The approximation of symmetric X-ray peaks by Pearson type VII distributions. *J. Appl. Cryst.*, 10:66–68 (1977). doi:10.1107/S0021889877012849. (page 202).

[187] S. Nukiyama and Y. Tanasawa. Experiments on the atomization of liquids in an air stream. Report 3 : On the droplet-size distribution in an atomized jet. *Trans. SOC. Mech. Eng. Jpn.*, 5:62–67 (1939). Translated by E. Hope, Defence Research Board, Ottawa, Canada. (page 202).

[188] P. Rosin and E. Rammler. The laws governing the fineness of powdered coal. *J. Inst. Fuel*, 7:29–36 (1933). (page 203).

[189] J. Laherrère and D. Sornette. Stretched exponential distributions in nature and economy: "fat tails" with characteristic scales. *Eur. Phys. J. B*, 2:525–539 (1998). doi:10.1007/s100510050276. (page 204).

INDEX OF DISTRIBUTIONS

Distribution	Synonym or Equation
β	beta
β'	beta prime
χ	chi
χ^2	chi-square
Γ	gamma
Λ	log-normal [179]
Φ	standard normal
Amaroso	(11.1)
anchored Amaroso	Stacy [1]
anchored exponential	See exponential (2.1)
anchored log-normal	See log-normal (6.1)
anti-log-normal	log-normal
arcsine	(12.6)
Appell Beta	(20.17) [132]
ascending wedge	See wedge (5.4)
ballasted Pareto	Lomax
Bates	(21.1)
bell curve	normal
beta	(12.1)
beta, J shaped	See beta (12.1)
beta, U shaped	See beta (12.1)
beta-exponential	(14.1)
beta-Fisher-Tippett	(21.2)
beta-k	Dagum [51]
beta-kappa	Dagum [51]
beta-logistic	(15.1) [1]
beta-log-logistic	generalized beta-prime [1]
beta type I	beta

---

†† Citations in this table document the origin (or early usage) of the distribution name.

SUBJECT INDEX

This guide is inevitably incomplete, inaccurate, and otherwise imperfect — *caveat emptor.*

www.ingramcontent.com/pod-product-compliance
Lightning Source LLC
Chambersburg PA
CBHW050602190326
41458CB00007B/2149